A PRACTICAL GUIDE TO AIRPLANE PERFORMANCE AND DESIGN

A PRACTICAL GUIDE TO AIRPLANE PERFORMANCE AND DESIGN

by

Donald R. Crawford

Published to help
the beginning designer
through some of the
preliminary phases of
aircraft design.

Third Printing, Revised

Crawford Aviation
1994

Copyright © 1979, 1981

by

Donald R. Crawford

All rights reserved, including that of translation into foreign languages; no part of this book or accompanying nomogram may be reproduced in any form without the specific permission of the author and publisher except in the case of brief quotations embodied in critical articles and reviews.

Library of Congress Catalog Card No. 81-67801
ISBN 0-9603934-0-4

Published by

Crawford Aviation
P. O. Box 1262
Torrance, Ca 90505

Printed in the United States of America
by
Delta Lithograph Company
28210 Avenue Stanford • Valencia, CA 91355-1111

Typeset by Dee Dee D'Arcy
Graphic Dimensions
15826 Hawthorne Boulevard, Suite A
Lawndale, California 90260

Photo by Jack Hobart

ABOUT THE AUTHOR:

Donald R. Crawford is an aircraft owner and a private pilot working on his instrument rating. He is also a hang-glider pilot and a member of the Experimental Aircraft Association, constructing a homebuilt aircraft with partners from the Chapter 40 Design Group. Currently, he is a member of the technical staff at TRW and a part-time lecturer in Aerodynamics at California State University, Long Beach. He received his Ph.D. degree from the University of California at Berkeley in Aeronautical Sciences. His wife, Sharon, is also a pilot and is active with the Long Beach Chapter of the Ninety Nines — the International Organization of Women Pilots.

To my air-lovin' uncle Wayne

and

my wife, Sharon (Sky Queen)

Preface

The nomogram and design procedure described in this book are aids I developed as a result of a design study for "Crawdad" — an ultralight foot-launched motor glider that I hope to build. In the process of learning about aircraft design I discovered the various useful performance and design nomograms of Raoul Hoffman, a noted practical aerodynamicist of the 30's and 40's. His alignment charts were useful because I could obtain answers to involved problems in aerodynamics by drawing straight lines connecting the related aircraft performance parameters. With this background, I was inspired to seek a single nomogram that I could use to make parametric studies for future designs. With the aerodynamic relations contained on a single graph, the calculation could then proceed in a systematic manner, since the answers from one graphical construction would be automatically available as data for the next.

An attractive feature of the nomogram is that it is "preprogrammed" so that we don't have to worry about exponents and conversion factors in the equations. All of the governing aerodynamic relations are built into the design of the scales and we can dispense with the tedious algebra.

Finally, by using a plastic template and a graphical construction technique, we can simplify the task of finding the rate of climb versus airspeed. One part of the template is used to find the gliding sink rate and the other part gives the rate of climb produced by the engine brake horsepower lifting the weight of the airplane with an efficiency corresponding to that of an idealized propeller. Using each part of the template as a French curve having the mathematically correct shape, we can immediately draw the two curves and subtract to find the rate of climb. The intersection of the curves gives the maximum level speed. The template is positioned using the reference points obtained from the graphical calculation with the nomogram. Then, we can use the reference scales on the template to make rapid parametric studies for the effects of altitude, weight, drag, brake horsepower, propeller diameter, and airplane size.

The book began as an instruction pamphlet for the use of the nomogram, but it has evolved into the present handbook, with the detailed sample calculation, parametric study and theoretical background. The governing equations are tabulated for the convenience of those who want to program them on a computer and a sample listing for a FORTRAN computer program for the performance analysis is given in Appendix E. However, the answers obtained from the nomogram and the graphical construction of the rate of climb curve are adequate in the preliminary design stages. Although the computer will give greater numerical accuracy, the approximations of the governing equations do not justify keeping accuracy greater than two or three digits.

The book is written for the homebuilder as well as for the beginning aerodynamicist. The first part of the book is devoted to practical considerations and is meant to be self-contained. The Airplane Performance and Design Nomogram can be used without really having to understand the details of the aerodynamic analysis of Part 2, which can be deferred until a working knowledge of the design process is established. In this way, the homebuilder or novice engineering student can gain confidence and experience by working with practical examples before trying to study the mathematics of aerodynamics. The book and nomogram will be especially useful as a supplementary text to aid the student with his (or her) preliminary aircraft design project.

I would like to acknowledge the encouragement of a number of my friends who also helped with a review of the rough-drafted material. These include Jerry Eakin, Jim Eninger, Jack Hobart, Vance Jaqua, Frank Kelly, Professor Larry Redekopp, my uncle Wayne Streeter, Dr. Norm Thompson, and Bob Young. They provided valuable comments and the feedback necessary for me to complete the text. I also want to thank Cheryl White for performing the task of typing the original manuscript; Lee O'Malley for helping with the technical editing; Carolyn and Neil for letting me be a part-time papa; and Sharon — my flying partner and bride — for her patience while I struggled with my Oshkosh deadline.

Donald R. Crawford

Redondo Beach, California
April 1979

Preface to the Revised Printing

The first edition of this book was introduced at the Experimental Aircraft Association Convention at Oshkosh, Wisconson, in 1979, and has been favorably received by students of Aerodynamics and builders of sport aircraft for the past two years. This revised printing is basically a corrected version of the first edition, with the collected errata incorporated into the text. Appendix L has been expanded to include performance data for some more recent airplanes. Appendix M has been written to show how one can determine the drag area and airplane efficiency factor from flight test data. Titles of some recent books and articles on specialized topics in Aerodynamics, as well as some classical texts that were overlooked in the first printing, have been added to the References.

I would like to thank the readers and reviewers for suggested changes in the presentation of the material.

May, 1981
Donald R. Crawford

Preface to the Third Printing, Revised

A few minor typographical errors have been corrected, but the majority of the material has been left unchanged. I have decided to leave the BASIC computer program as an item separate from the book, since there is a good chance that it will evolve further. I would like to thank the steady support of my readers.

October, 1994
Donald R. Crawford

CONTENTS

Preface ix

Introduction 1

Part 1. PRACTICAL APPLICATIONS

Descriptive Design Procedure 5

Worked Example — Thorp T-18 Tiger 9

Relations for the Airplane Performance and Design Nomogram 13

① $C_L, V, W/S$ — Lift Coefficient, Airspeed, Wing Loading 15

    Stall Analysis, Airfoil Selection Process 16

② $S, W/S, W$ — Wing Area, Wing Loading, Gross Weight 19

③ S, b_e, eAR, c_e — Wing Area, Effective Span, Effective Aspect Ratio, Effective Chord 21

④ $b_e, W/b_e, W$ — Effective Span, Effective Span Loading, Gross Weight 25

⑤ A_D, V_{max}, THP_a — Drag Area, Maximum Level Speed, Available Thrust Horsepower 27

⑥ $C_{D,O}, A_D, S$ — Zero-Lift Drag Coefficient, Drag Area, Wing area 31

⑦ $A_D, V_{minS}, W/b_e, THP_{min}, D_{min}$ — Drag Area, Airspeed for Minimum Sink,

    Effective Span Loading, Minimum Power Required for Level Flight,

    Minimum Drag 33

⑧ $R_{S,min}, THP_{min}, W$ — Minimum Sink Rate, Minimum Power Required

    for Level Flight, Weight 39

⑨ $A_D, b_e, (L/D)_{max}$ — Drag Area, Effective Span, Maximum Lift-to-Drag Ratio 43

⑩ $A_D, C_{L,minS}, c_e$ — Drag Area, Lift Coefficient at Minimum Sink, Effective Chord 45

⑪ $W, BHP, R^*_{C,max}$ — Weight, Engine Brake Horsepower, Ideal Maximum Rate of Climb 49

⑫ T_s, BHP, V_{prop}, D_p — Static Thrust, Engine Brake Horsepower, Reference

    Propeller Airspeed for 74% Efficiency, Propeller Diameter 51

⑬ D_p, RPM, M_p — Propeller Diameter, Propeller Rotational Speed, Propeller Tip

    Mach Number 57

⑭ V_{minS}, $R_{S,min}$, $(L/D)_{max}$ — Airspeed for Minimum Sink Rate, Minimum Sink Rate, Maximum Lift-to-Drag 61

⑮ $C_{D,O}$, $C_{L,minS}$, eAR, $(L/D)_{max}$ — Zero-Lift Drag Coefficient, Lift Coefficient at Minimum Sink, Effective Aspect Ratio, Maximum Lift-to-Drag Ratio 65

⑯ $(L/D)_{max}$, D_{min}, W — Maximum Lift-to-Drag Ratio, Minimum Drag, Gross Weight 69

Summary of the Nomogram Calculation for the T-18 71

How to Use the Template for Calculation of Rate of Climb 75

Parametric Study of the Baseline Design 82

    Power Setting, Cruise Speed, Range 82

    Altitude Effects; Absolute and Service Ceilings 85

    Weight Effects 92

    Drag Effects: Streamlining and Flaps 93

    Effective Span Changes; Ground Effect 95

    Power Effects: Larger Engine, Twin vs. Single Engine, Propeller Diameter 96

    Comparison of the Parametric Variations of Weight, Drag, Span, and Power 97

Sample Calculations of Aircraft Performance 99

    Gossamer Condor and Gossamer Albatross — Man-Powered Aircraft 99

    Powered Quicksilver Hang Glider --- Microlight 102

    Cessna 172 — General Aviation Aircraft 105

    "Crawdad" — Foot-Launched Motorglider Design Study 108

Airplane Performance Comparisons 115

Part 2. THEORETICAL BACKGROUND

Low Speed Aerodynamics 120

    Force Balance in Gliding Flight 120

    Induced Drag 123

    Minimum Sink Rate 125

Maximum Lift-to-Drag Ratio	128
Level Flight	129
Climbing Flight	130
Idealized Propeller Theory	131
Propeller Efficiency	132
Advance Ratio, Power Coefficient, and Nondimensional Velocity $\tilde{V}$	135
Static Thrust	136
Propeller Tip Speed	138
Summary of Performance Relations	138

APPENDIX

Appendix A.	Abbreviations and Symbols	143
Appendix B.	What is a Nomogram?	149
Appendix C.	Discussion of Units	153
Appendix D.	Standard Atmosphere	155
Appendix E.	FORTRAN Computer Program for Performance Analysis	159
Appendix F.	Airplane Efficiency Factor, e; Ground Effect	165
Appendix G.	Drag Analysis	169
Appendix H.	Airfoil Selection	177
Appendix I.	Reynolds Number, $Re = \rho V \ell / \mu$	183
Appendix J.	Equation of State, $p = \rho R T$	185
Appendix K.	How to Find the Solution of a Cubic Equation	187
Appendix L.	Tabulated Performance Data for Various Aircraft	189
Appendix M.	How to Calculate Drag Area, A_D, and Efficiency Factor, e, from Flight Test Data	182*
References and Further Reading		203, 188*
Index		205

*Added in Revised Printing, 1981

INTRODUCTION

The first question the airplane enthusiast usually asks is, "How fast does it do?" followed by, "What's the stall speed? How much horsepower does the engine have? What is the best rate of climb? Useful load? Gross weight? . . ." By finding the answers to these questions, he can compare the performance of one airplane with another. The designer asks himself these same questions, but from a slightly different point of view. He wants to know how to make the airplane go faster, stall slower, climb faster and carry more load within the limitations of the available engine size, airplane gross weight and pocketbook.

The approximate analyses described in this book can be used to help the first-time designer calculate the complete performance of the airplane from seven basic parameters: gross weight, drag area, wing span, wing area, maximum lift coefficient of the airfoil, engine horsepower, and propeller diameter. All airplane performance and design variables are mathematically related to these seven parameters by equations derived in Part 2 and through the Airplane Performance and Design Nomogram described in Part 1 of this book. Most preliminary design problems can be solved by drawing straight lines on the nomogram. The key quantities that come from the graphical calculations are:

1. $R_{S,min}$ — The idealized minimum sink rate

2. V_{minS} — The corresponding airspeed for minimum sink rate

3. $(L/D)_{max}$ — The maximum lift-to-drag ratio

4. V_{prop} — The airspeed which gives idealized propeller efficiency of 74%

5. $R^*_{C,max}$ — The rate of climb produced if all the brake horsepower of the engine were used to lift the weight of the airplane

6. T_s — The static thrust available from the idealized engine-propeller combination.

With these parameters and the dimensionless rate-of-climb and sink-rate curves (which are also included on a plastic template), the rate of climb can be found as a function of the airspeed. After the baseline design is determined, we can easily use the template to find the parametric effects of altitude, weight, drag, span, power, and propeller diameter on the airplane performance.

The first part of the book is organized as a handbook that can be used without any higher mathematics. The Airplane Performance and Design Nomogram is broken down into its elements and described separately in a step-by-step fashion. The use of each relation is illustrated by working through a sample design calculation based on the popular Thorp T-18, a modern two-place sport aircraft. The results of a study of this design show the effects of drag reduction, weight reduction, flaps, power setting, and turbocharging on overall performance. Then, a performance rating parameter and a kinetic energy parameter are used to compare various types of aircraft.

Part 2 describes the theoretical aerodynamics of low speed flight and is more mathematically oriented. The equations derived there form the basis for the construction of the nomograms and dimensionless figures used in Part 1. First, we discuss the aerodynamics of equilibrium gliding, level, and climbing flight. Then, we develop the idealized propeller theory.

In the Appendix, we have gathered some supplementary sections that are useful in the course of the design process. These include discussions of drag area, airplane efficiency factor, airfoil selection, Reynolds number, equation of state, and others. Together with Parts 1 and 2, we have a concise description of airplane performance that can be useful to the homebuilder and beginning aerodynamicist.

PART 1. PRACTICAL APPLICATIONS.

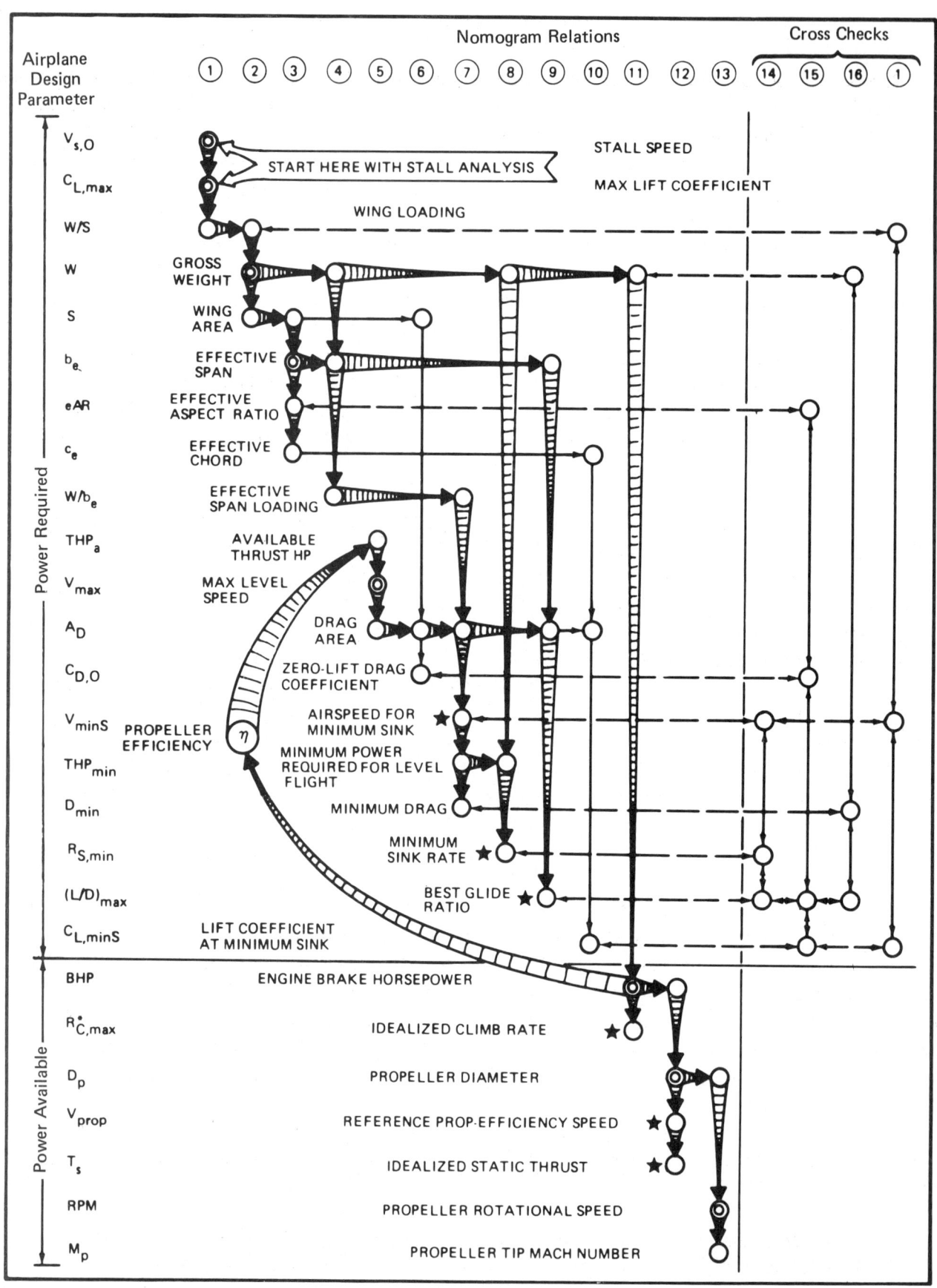

Figure 1. Flow Chart for Design Procedure.

DESCRIPTIVE DESIGN PROCEDURE

Before we describe how to make the step-by-step detailed calculations with the Airplane Performance and Design Nomogram and the related template, we will briefly outline one method for airplane design, represented graphically on Figure 1. The numbered relations listed at the top of the figure — ①,②,③, ... — refer to the way in which the various airplane performance parameters are dependent on each other. These parameters are listed down the left-hand side of the figure and will be defined in the course of the design process. (Also see definitions in the Appendix.) The object of the flow diagram is to point out how to determine the most important parameters for airplane performance: V_{minS}, idealized value for the airspeed at minimum sink conditions; $R_{S,min}$, the minimum sink rate; V_{prop}, the reference speed corresponding to a certain value of the ideal propeller efficiency; R_C^*, the climb rate that would be obtained if the brake horsepower of the engine were used to lift a weight equal to that of the airplane; $(L/D)_{max}$, the maximum lift-to-drag ratio; and T_s, the idealized static thrust. These quantities are emphasized in the figure because they will be used later with the template to find the rate of climb versus airspeed. We see that there are several paths that could be used to find all of the variables. For the sample design procedure, the relations on the right hand side of the figure are redundant and can be used for cross checks of the design method.

In order to begin the design procedure, we start in the upper left-hand corner and choose a desired stall speed, $V_{s,O}$, and an estimate of the maximum lift coefficient for the airfoil, $C_{L,max}$. We also have to decide whether we intend to use flaps, and choose the speed and lift coefficient accordingly. With these quantities chosen and using relation ①, we can draw straight lines on the nomogram connecting $V_{s,O}$, $C_{L,max}$ and W/S to determine the wing loading. This wing loading satisfies the stall condition and determines the ratio of the gross weight of the airplane to its wing area, in pounds per square foot (lb/ft^2). If the wing loading goes up, so will the stall speed, unless we pick a better airfoil, or add flaps or other high lift devices to the wing to increase the maximum lift coefficient. An airplane with a wing loading of 10 lb/ft^2 and a flaps-up stall speed of 57 mph will have a maximum lift coefficient of about 1.2 (which is a typical value). Relation ① is an expression of the force balance in the vertical direction (as discussed in the theoretical section). As such, it applies to other flying speeds as well as the stall condition. That is, for a given wing loading, we can use the nomogram to determine the lift coefficient if we know the flying speed.

On Figure 1, note that we now have a value to be used in relation ② that came from the solution of the relation ① — W/S, the wing loading. To proceed we need to estimate the gross weight of the airplane. This can be done by looking at airplanes that are similar to the design we have in mind. The gross weight is not an easy quantity to estimate in the early phases of the design, because there are so many unknown weights: payload, fuel, powerplant, and structure. At this stage it is a good idea to choose the engine that we plan to use to meet our design goals: maximum level speed, maximum rate of climb, range, etc. If the engine is chosen now we will have one design point pinned down — maximum available power — and we will be better able to make an estimate of the final gross weight. Then, using relation ②, we can find the required wing area from the knowledge of the wing loading and the weight. This could have been done by division, but one purpose of the nomogram is to eliminate as many of the mathematical steps as possible, including division.

We have established a trend: whenever there is a relation that has two known variables, we can determine the other(s) by drawing a straight line on the nomogram and extending the line to intersect the axis of the unknown parameter(s) in the relation. In the case of relation ①, we found the wing loading, W/S, from the stall speed, $V_{s,0}$, and the maximum lift coefficient, $C_{L,max}$. In relation ②, we found the wing area, S, from the wing loading, W/S, and the weight, W. In this manner, we can continue down and across the flow chart, where we have to assume the value of a parameter when we know only one of the parameters in a particular relation. For instance, in relation ③ we have to choose a value for the wing span, b, since we only know the wing area, S. Equivalently, we could have chosen the aspect ratio, AR, or the mean chord, c, since all of these quantities are geometrically related to each other. Note that b, AR, and c are modified by an efficiency factor e. This will be discussed later in the section describing relation ③ and in the supplementary section on airplane efficiency factor.

The effective span loading, W/b_e, is one of the basic parameters for finding the power required for level flight (relation ⑦) and can be found from the previously assumed values for weight and effective span, using relation ④. Whereas the wing area was chosen to satisfy the stall requirements, the span is chosen large enough to obtain a reasonable rate of climb. Airplanes with larger spans will normally have better climb characteristics than airplanes with the same weight and wing area, but with a smaller span.

Before we can make further progress toward finding the sink rate, we need to determine the drag area of the airplane, A_D. The drag area is defined to be the fictitious area that, when placed normal to the direction of the airstream, would have the same drag as the entire airplane. Typical values for drag area range from 1-2 square feet for small streamlined designs, to more than 20 ft^2 for large "dirty" airplanes. The way to determine the drag area is discussed in Appendix G located at the end of the book. In order to decide how "clean" — aerodynamically speaking — to make the design, we need to ask ourselves how fast we want to go: V_{max}, the maximum level flight speed. We also need to estimate how much of our original brake horsepower is left for useful work against the drag forces after we accelerate the air through the propeller to produce thrust. If we assume that we have 80 percent of the engine brake horsepower, BHP, available as thrust horsepower, THP_a, we can use relation ⑤ to find the maximum allowable drag area for the airplane that will let us meet our high speed requirement. Assuming a value of 80 percent for the propeller efficiency is good first approximation that will be refined later. If we can reduce the drag area below this value, we will be able to have a higher maximum level speed. Since the drag area is some fraction of the projected frontal area of the airplane, it makes sense to keep the frontal area as small as possible to have good high speed performance. Also, because the viscous forces act on the surface of the airplane, we should try to keep the wetted area of the airplane as small as possible. Since the wing area has already been set by the stall requirements, the drag analysis will emphasize fuselage streamlining, engine cooling drag, and the selection of airfoil sections with low drag — unless we also want to sacrifice climb performance for high speed performance by making short wings (like the clipped-wing airplanes seen in air races).

Although the zero-life drag coefficient, $C_{D,O}$, may seem like a parameter of fundamental importance, we can see from the flow chart in Figure 1 that drag area is the key link between the maximum level speed, V_{max}, and the speed for minimum sink rate, V_{minS}. The drag coefficient, $C_{D,O}$, found from relation ⑥ is the ratio between the drag area, A_D, and the wing planform area, S. The key relation, however, turns out to be relation ⑦ where the effective span loading and the drag area determine: (1) the speed for minimum sink rate; (2) the minimum thrust horsepower required for level flight, THP_{min}; and (3) the minimum drag, D_{min}. The speed for minimum sink rate is an idealized quantity and may turn out to be smaller than the stall speed. This is no problem, however, and will be taken into account when we draw the rate of climb curves as we will see later. Since the flying speed cannot be smaller than the stall speed, a graphical adjustment to the sink rate will be made. This will be discussed in more detail in the section describing R_S vs V — sink rate versus airspeed. We can find the minimum sink rate, $R_{S,min}$, using relation ⑧, weight, W, and the minimum power required for level flight, THP_{min}.

The lift-to-drag ratio is a measure of how far a gliding airplane will travel as it descends. A glider with a lift-to-drag ratio of 20 will travel forward 20 feet for each foot of altitude that it loses. The maximum lift-to-drag ratio, $(L/D)_{max}$, is therefore a very important performance parameter for gliders. We can find $(L/D)_{max}$ from the effective span, b_e, and the drag area, A_D, using relation ⑨. It is important to note that the maximum lift-to-drag ratio does not depend on the weight of the airplane. A heavier airplane with the same streamlined shape will have the same lift-to-drag ratio, but will fly at a faster airspeed. A jumbo jet has a good $(L/D)_{max}$ but it would not make a good sailplane because of the high sink rate. An airplane with a good lift-to-drag ratio will have a large effective span and a small drag area, as in the modern sailplanes.

Finally, the lift coefficient of the wing under minimum sink conditions $C_{L,minS}$, can be found using relation ⑩, the drag area, A_D, and the effective chord, c_e. There is no need to panic if the lift coefficient — $C_{L,minS}$ — turns out to be larger than the maximum lift coefficient. This will alert us and help to remind us that the minimum sink conditions just calculated have to be modified to account for the actual stall condition. The theoretical aspects of this are discussed in Part 2 where we talk about the induced drag and the dependence of the induced drag coefficient on the square of the lift coefficient (the so-called drag polar). The practical consequences are explained in the section on sink rate versus airspeed (R_S vs V).

We have now described all of the performance parameters that are related to the power-required side of the ledger. Before we discuss the variables concerning the power available, we can look at the cross-check features of the nomogram. These are the relations on the right hand side of the flow chart — relations ⑭, ⑮, ⑯ and ①. First, since the sink rate and the airspeed determine a glide angle, and since the glide angle at minimum sink is related to the glide angle under the best L/D conditions, we can relate $R_{S,min}$, V_{minS}, and $(L/D)_{max}$ through relation ⑭. If it turns out that this relation is not satisfied, there is something wrong and the errors must be found before the graphical construction can continue. Similarly, the effective aspect ratio, eAR, the zero-lift drag coefficient, $C_{D,O}$, lift-to-drag ratio, $(L/D)_{max}$, and the lift coefficient at minimum sink, $C_{L,minS}$, are dependent on each other through relation ⑮. The maximum lift-to-drag ratio is obviously related to the weight and the minimum drag, since the weight is equal to the lift in equilibrium flight. This is the statement made by relation ⑯. And, finally, relation ① is an expression of the force balance in the vertical direction and applies to all flight conditions: stall, minimum sink, best L/D, maximum speed, etc.

The power-required side of the performance equation is contained on the bottom half of the flow diagram in Figure 1. We want to find the idealized climb rate, $R^*_{C,max}$, reference speed for propeller efficiency, V_{prop}, and static thrust, T_s. The idealized climb rate is defined by the rate at which a weight equal to that of the airplane would be lifted if all of the brake horsepower of the engine were used in the process. This climb rate is proportional to the power-to-weight ratio that is often talked about in performance discussions. Relation ⑪ shows the key part that weight plays in relation to the rest of the performance parameters. The idealized maximum rate of climb, $R^*_{C,max}$, will be used to locate the horizontal axis of the template for the ideal powered climb rate in the same way that the minimum sink rate was used with the overlay for velocity dependence of the gliding sink rate. The other parameter needed for the location of the climb-rate template is the reference propeller efficiency speed, V_{prop}. This parameter comes from idealized propeller theory and is related to the propeller diameter and engine brake horsepower — shown in relation ⑫. The idealized static thrust can also be calculated from this relation. The limitations of the approximate propeller theory are discussed in the section where we calculate R^*_C vs V. Further comments are made in the detailed derivation found in Part 2.

7

Finally, relation ⑬ gives the propeller tip Mach number if the rotational speed and propeller diameter are known. This is used mainly as a check, since we want to keep the propeller tip Mach number less than about 0.8 for better efficiency and less noise. Note that the rotational speed refers to the propeller rotational speed, not to the engine shaft speed. This way we can pick a large propeller diameter and choose the gear ratio that will keep the propeller tip speeds low.

In the next few sections, each of the relations will be described and a sample calculation based on the Thorp T-18 will illustrate the use of the elements on the nomogram. Then, all of the elements will be combined to show the geometric interrelation of the lines constructed on the nomogram. If there is some confusion in the course of the geometric solution on the nomogram, these individual sections may be referred to.

After all of the nomogram relations are described, we will discuss the construction of the idealized rate of climb (for the engine lifting a weight with an efficiency corresponding to the appropriate propeller efficiency) and the idealized sink rate, including modifications to take into account the actual stall condition. Then, we can find the rate of climb by subtracting these rates. The level flight speed will occur when the two curves cross. We will mention here that the calculations are first performed for our design at sea-level. Then, we can use the templates to find the performance at altitude and other parametric studies.

WORKED EXAMPLE — Thorp T-18 Tiger

The Thorp T-18 was the example selected for calculation because it is representative of the type of aircraft that a homebuilder may want to build or modify. Figure 2 is a picture of the T-18 and Figure 3 is a three-view with tabulated performance data. The T-18 is a high performance sport aircraft with two-seat side-by-side seating. The all-metal cantilever low-wing monoplane was designed for engines in the 108-200 hp category. The landing gear is fixed and streamlined to reduce the drag of the airplane. The T-18 is capable of 200 mph in level cruising flight. A T-18 built by Don Taylor, Victoria '76, was the first homebuilt aircraft to fly around the world — 26,200 miles over a period of 61 days in the summer of 1976.

In the design example, we will show how to begin with the desired performance and find the airplane parameters that will satisfy these requirements. We will sometimes work backwards from the known values for the T-18 to show that the airplane does indeed satisfy the relations on the nomogram and the dimensionless curves on the template. Then, we will show how to make a parametric study to see how sensitive the performance characteristics are to each of the design variables.

Figure 2. Thorp T-18, Built and Flown by Lu Sunderland.

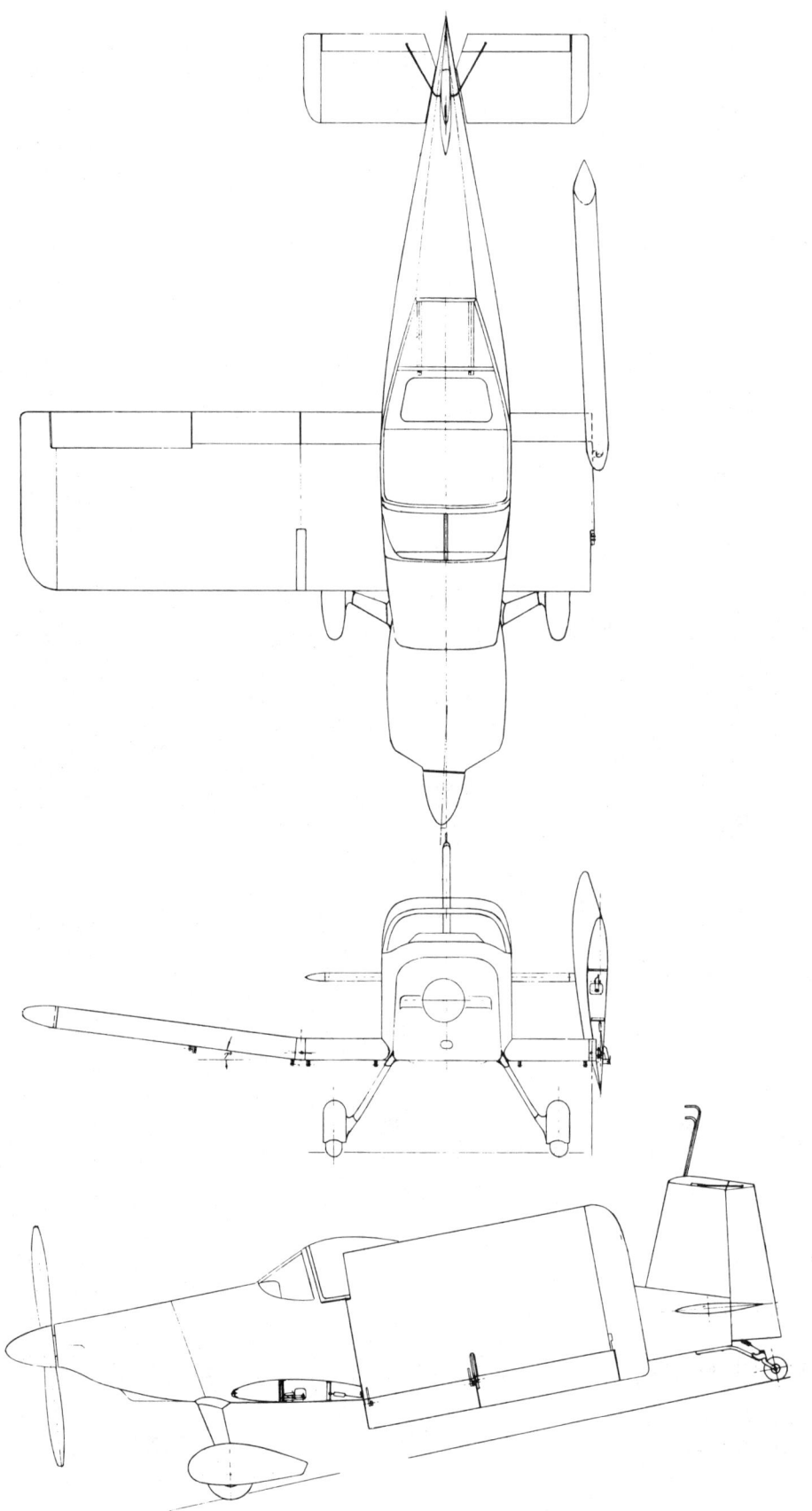

Figure 3. Convertible Wing Thorp T-18C Three-View (Sunderland Aircraft)

Thorp T-18 Performance Data

Wing Span	b = 20 ft 10 in
Wing Chord, Constant	c = 4 ft 2 in
Wing Area,	S = 86 ft^2
Length Overall	18 ft 2 in
Height Overall	4 ft 10 in
Tailplane Span	6 ft 11 in
Propeller Diameter	D_p = 63 in
Weight Empty	W_e = 900 lb
Max Take Off Weight	W = 1506 lb
Engine, Lycoming	BHP = 180 hp
Max Level Speed at SL	V_{max} = 200 mph
Max Cruising Speed	V_c = 175 mph
Stalling Speed	$V_{s,0}$ = 65 mph
Max rate of climb	$R^*_{C,max'}$ = 2000 ft/min
Service Ceiling	20000 ft
Take Off Run	300 ft
Landing Run	900 ft
Range with Max Fuel	R = 500 miles
Useful Load	W_u = 606 lb
*Performance Rating	F_p = 0.138
*Kinetic Energy Parameter	WV^2_{max} = 6.02 x 10^7 lb mph^2
*Drag Area/Propeller Efficiency	A_D/η = 3.3 ft$^2 \longrightarrow A_D \approx$ 2.6 ft^2

*Calculated. See Appendix L.

RELATIONS FOR THE AIRPLANE PERFORMANCE AND DESIGN NOMOGRAM

How to use the Airplane Performance and Design Nomogram:

1. Find a relation that has two known quantities.

2. Draw a line on the nomogram connecting the known parameters with the other variable(s) in the relation.

3. The new value(s) are now available for use as input for the next relation. Repeat Step 1.

4. Continue the graphical calculation until $R_{S,min}$, V_{minS}, V_{prop}, and $R^*_{C,max}$ are determined. These parameters serve as reference points to be used with the plastic template to determine the airplane climb performance.

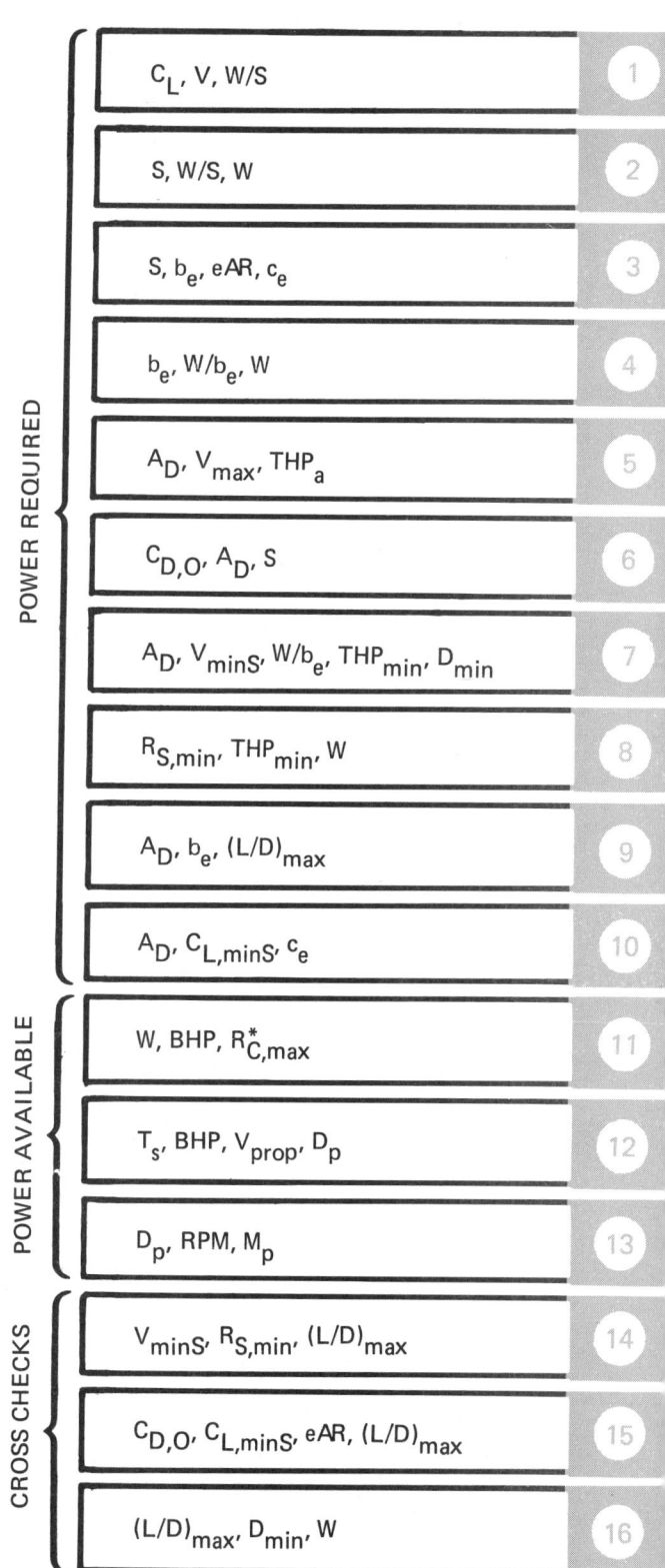

POWER REQUIRED:
1. C_L, V, W/S
2. S, W/S, W
3. S, b_e, eAR, c_e
4. b_e, W/b_e, W
5. A_D, V_{max}, THP_a
6. $C_{D,O}$, A_D, S
7. A_D, V_{minS}, W/b_e, THP_{min}, D_{min}
8. $R_{S,min}$, THP_{min}, W
9. A_D, b_e, $(L/D)_{max}$
10. A_D, $C_{L,minS}$, c_e

POWER AVAILABLE:
11. W, BHP, $R^*_{C,max}$
12. T_s, BHP, V_{prop}, D_p
13. D_p, RPM, M_p

CROSS CHECKS:
14. V_{minS}, $R_{S,min}$, $(L/D)_{max}$
15. $C_{D,O}$, $C_{L,minS}$, eAR, $(L/D)_{max}$
16. $(L/D)_{max}$, D_{min}, W

Relation ①: Lift Coefficient, Airspeed, Wing Loading

$C_L, V, W/S$ | 1

Relation ① is a form of the equilibrium force balance in the vertical direction (lift force = weight). The mathematical expression is given by $W/S = C_L V^2/391$. The relation is used for the stall analysis and for the airfoil selection process, where we find the lift coefficient at our desired flight conditions.

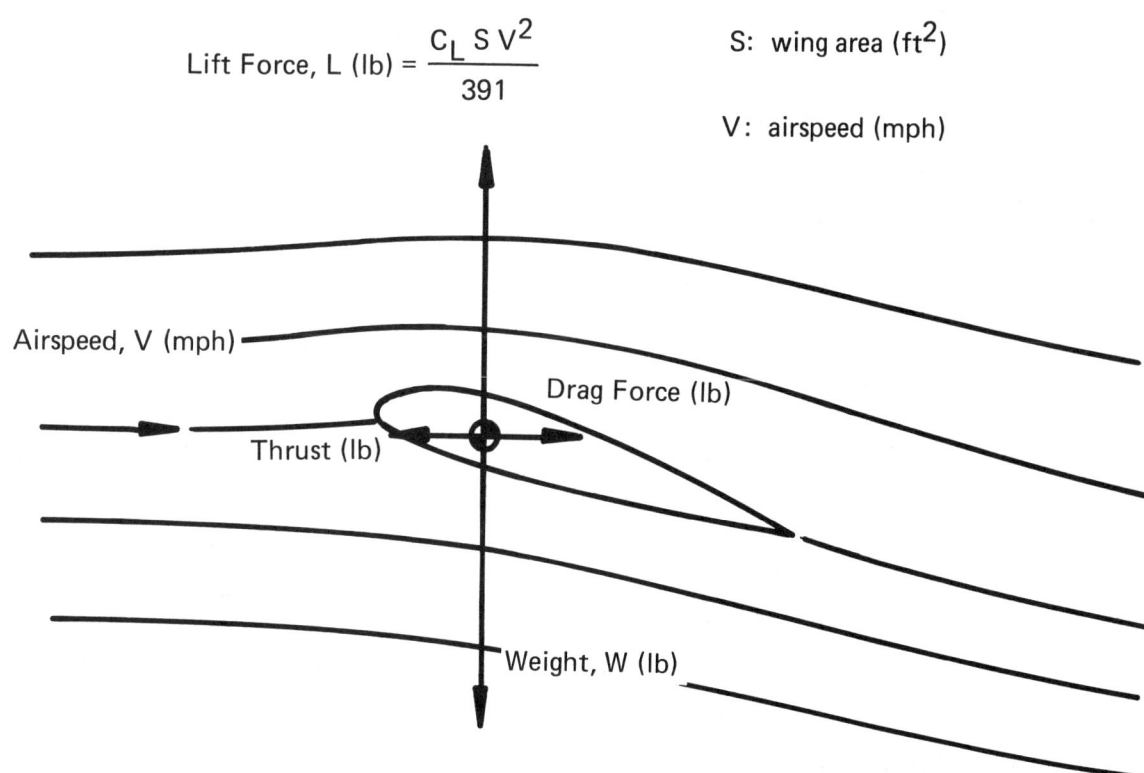

Lift Force, L (lb) = $\dfrac{C_L S V^2}{391}$

S: wing area (ft^2)

V: airspeed (mph)

STALL ANALYSIS

If we plan to have an airplane with as low a stall speed as possible and still have a small wing area, we need to select an airfoil with a large value for the maximum lift coefficient. Other factors enter into the airfoil selection process, including the pitching moment of the airfoil section, the minimum drag coefficient, and the shape of the curve of the lift coefficient versus angle of attack curve near the stall point. Details of the airfoil selection process are described to Appendix H.

We know that the Thorp T-18 has a NACA 63_1-412 (see Appendix H). Since this airfoil has a maximum lift coefficient of 1.52 without flaps, and we desire a stall speed of about 67 mph, we can find the wing loading from relation ① on the nomogram in Figure 4. Connecting the values for $C_{L,max}$ and $V_{s,O}$, we find that the wing loading, W/S, is 17.4 lb/ft^2. Increasing the value of $C_{L,max}$ will decrease the stall speed for the same wing loading.

AIRFOIL SELECTION PROCESS

Since relation ① is a general expression of the equilibrium force balance in the vertical direction, we can use it to find the lift coefficient that corresponds to a particular speed, if we have a known value for the wing loading. In the case of the T-18, where we have already calculated the wing loading that corresponds to the stall condition, we find from Figure 4 that the lift coefficient will be 0.21 at the maximum level speed of 180 mph. We can use this fact when we try to find the best low-drag airfoil at this lift coefficient with a Reynolds number that corresponds to this condition. (See Appendix I for a discussion of the Reynolds number.)

An alternative form for relation ① is presented in Figure 5 where the lift coefficient is plotted versus the airspeed for various values of wing loading. The wing loading can be determined by entering the graph with $C_{L,max}$ and the stall speed. Then, drawing a parallel line, find the lift coefficient at other speeds.

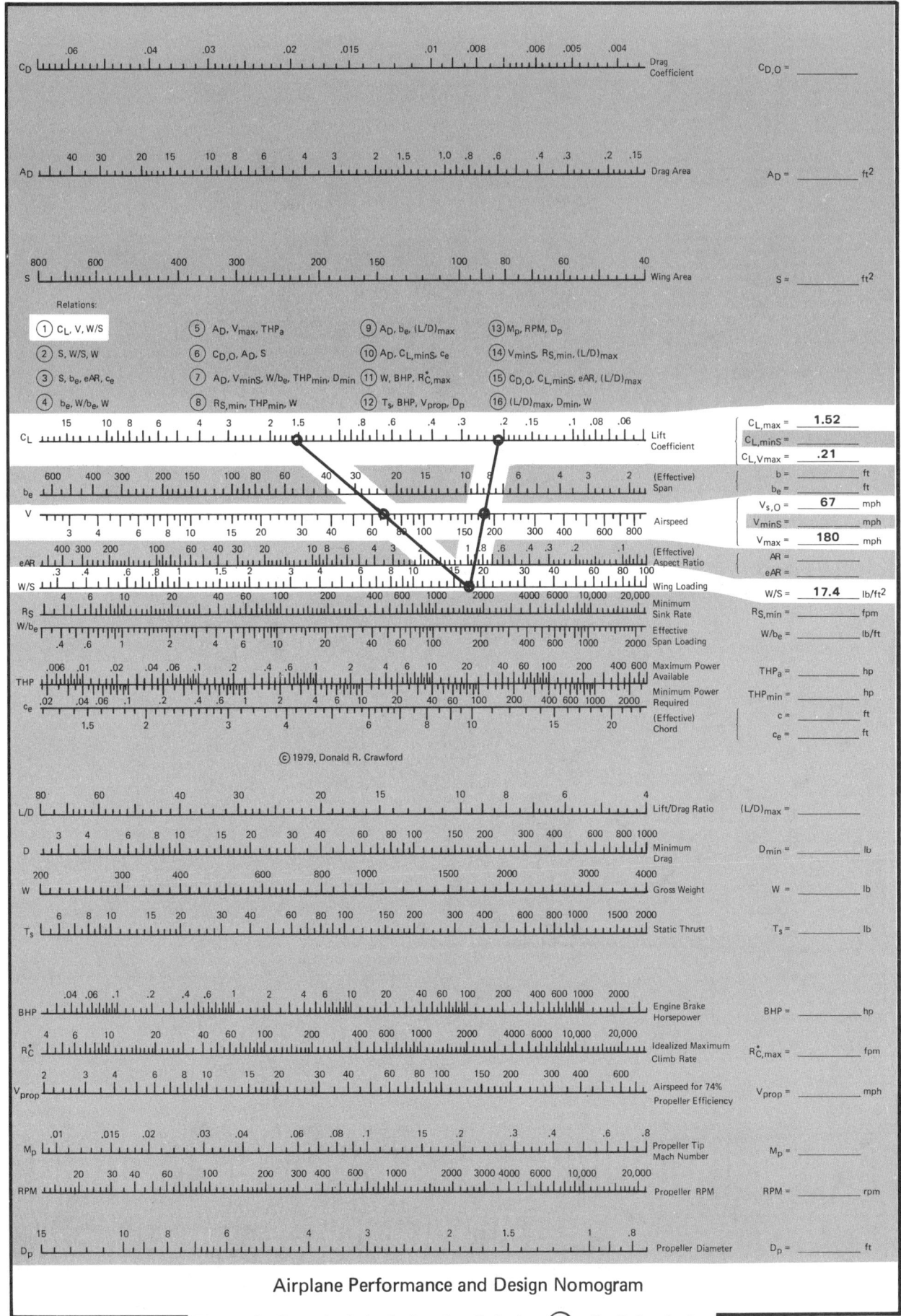

Airplane Performance and Design Nomogram

Figure 4. Sample Calculation for Relation ①. Stall Analysis.

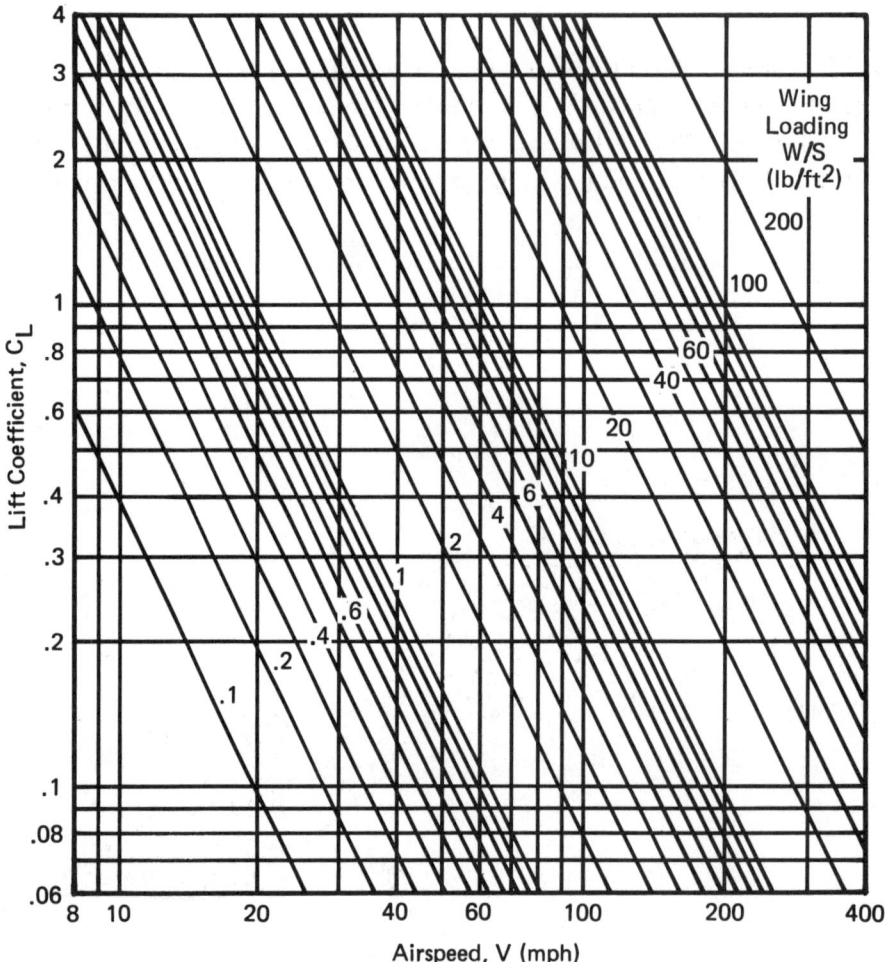

Figure 5. Relation ①: Lift Coefficient, Airspeed, Wing Loading. $W/S = C_L V^2/391$.

Relation ② : Wing Area, Wing Loading, Gross Weight

S, W/S, W	②

Relation ② is merely a definition for the wing loading. That is, the wing loading is defined as the gross weight divided by the wing area. The nomogram can be used to find the wing area if the weight and the wing loading are known. This is the case for the sample calculation in Figure 6 where we know the wing loading from the stall analysis and the weight from a preliminary weight estimate.

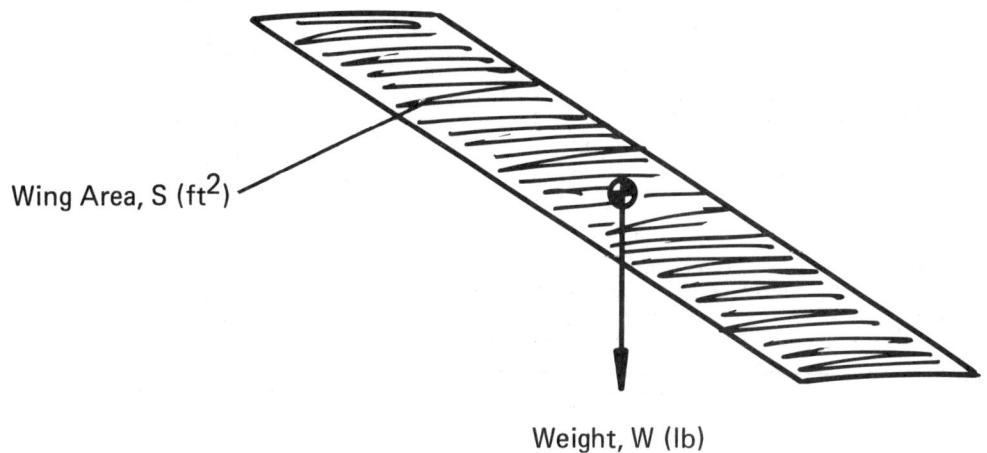

Wing Area, S (ft^2)

Weight, W (lb)

Wing Loading, W/S (lb/ft^2)

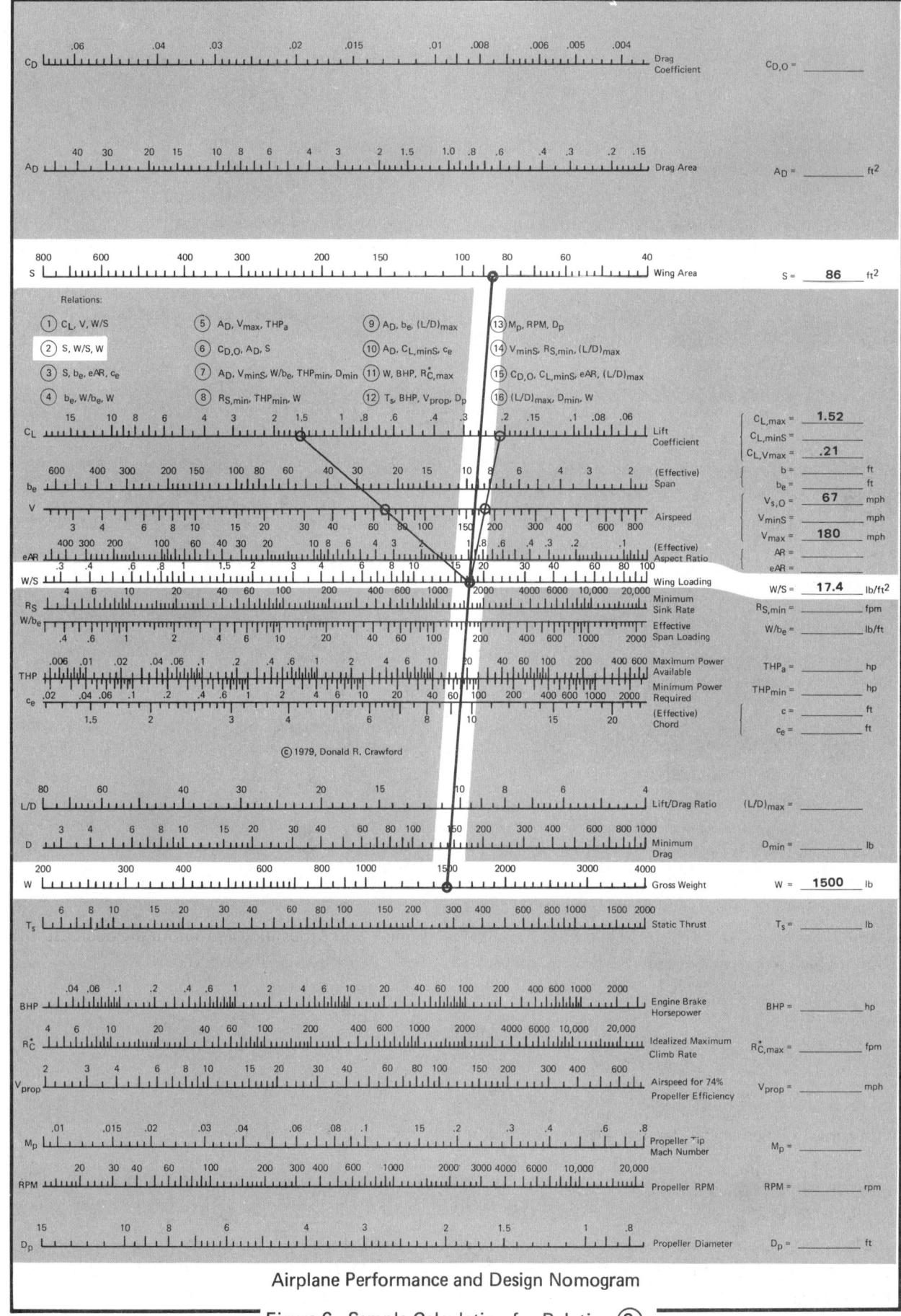

Figure 6. Sample Calculation for Relation ②.

Relation ③: Wing Area, Effective Span, Effective Aspect Ratio, Effective Chord

Relation ③ can be used with the geometric wing span, aspect ratio, and mean chord as well as the effective span, effective aspect ratio, and effective chord, which have been modified using the airplane efficiency factor, e. This relation is essentially a method of defining the (effective) aspect ratio and the (effective) chord if we know the wing area, S, and the (effective) span. The definition of average chord is the wing area divided by the span ($c = S/b$). The aspect ratio is then defined as the ratio of the span to the average chord ($AR = b/c$). If we substitute the definition for the chord from the above expression, we can determine the aspect ratio directly from the wing area and the span ($AR = b^2/S$). Since we have not mentioned what the planform shape is, these expressions hold for all wing shapes; rectangular, tapered, elliptical, delta, etc.

S, b_e, eAR, c_e ③

The effective aspect ratio, eAR, is a quantity that appears naturally in the discussion of induced drag (see Part 2). It is the geometric aspect ratio multiplied by an airplane efficiency factor, e. This e-factor has no real theoretical justification except that it allows the use of Prandtl's theory for a wing of finite span, if we use a fictitious wing with the equivalent aspect ratio, eAR. The method for determining the efficiency factor is given in Appendix F.

The airplane efficiency factor is a function of the geometric aspect ratio and the shape of the planform. According to Prandtl's theory, a wing with an elliptical lift distribution will give the least amount of induced drag for a given amount of lift. This is the reason for the choice for this planform for the Spitfire of World War II fame. Additional factors that affect the airplane efficiency factor are the shape and size of the fuselage compared to the wing, and the airfoil section used for the wing.

As an example, the four planforms in Figure 7 have the same geometric aspect ratio, ($AR = 4$) since they have the same wing area and the same span. The airplane efficiency factor changes the planforms into the equivalent shapes with the new effective aspect ratios and the new effective spans. In the calculations for the airplane performance, the effective span is one of the most important parameters for determining the sink rate (and the rate of climb for powered flight). Therefore, it is desirable to choose a planform that will give the largest effective span for the given amount of wing area. The rectangular and tapered planforms give results that are not too bad compared to the elliptic planform (and are a lot easier to construct). These examples were made neglecting the effects of the fuselage and other factors that might decrease the efficiency factor. When other factors are taken into account, the airplane efficiency factor for normal airplanes usually lies in the range 0.7 to 0.85. In ground effect — when the airplane is within an altitude of one wingspan above the surface — the airplane efficiency factor is greatly increased. This reduces the power required if the airplane is flown at the same airspeed and is one of the reasons the manpowered airplanes fly close to the ground.

As a sample calculation, let us determine the effective span for the T-18. First, choose a value for the geometric span (b = 20 ft 10 in). From the stall analysis we need a wing area of 86 ft^2. Draw a line on the nomogram in Figure 8 connecting the span and the wing area to find the aspect ratio ($AR = 5$) and the average chord ($c = 4.1$ ft). Using the method of Appendix F, find the value for the airplane efficiency factor ($e = 0.74$). Now, multiply the aspect ratio by this value by shifting the value for AR by the distance between 1. and 0.74 to obtain the effective aspect ratio ($eAR = 3.7$). Connect the effective aspect ratio with the wing area to find the effective span ($b_e = 18$ ft) and the effective chord ($c_e = 4.75$ ft).

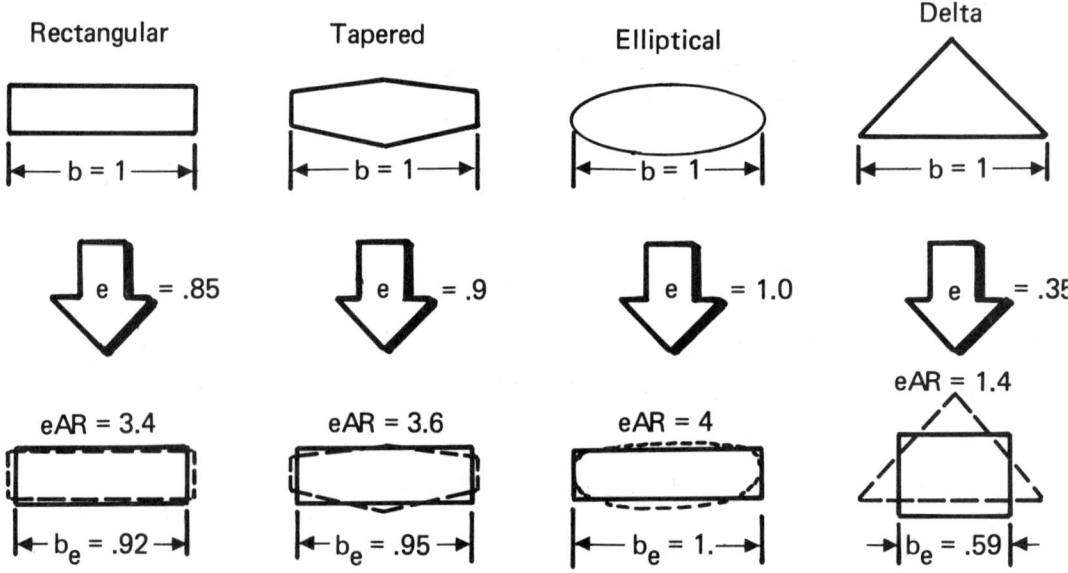

Figure 7. Effective Aspect Ratio for 4-Wings of Various Planform Shapes with the same Geometric Span and Wing Area.

Relation ③ can also be determined from Figure 9, where we have plotted the wing span, b, versus the chord for various values of wing area, S, or aspect ratio, AR. Knowing any two values will let us find a point on the figure which will then give us the other two values.

The span efficiency factor for biplanes and canards must be calculated in a different manner from that given in Appendix F. Since the aspect ratio, AR, is defined by span-squared divided by wing area (b^2/S), the efficiency factor for biplanes and canards will usually be greater than one (of the order 1.15). (The drag area will most likely be larger than a monoplane of the same span, since there is more wetted surface area for skin-friction drag). Details on how to calculate this factor for various values of span ratio, gap, stagger, etc. are given by Von Karman & Burgers (1935), Betz (1935), Millikan (1941) and Laitone (1978a, 1978b, 1979, 1980.) These methods use biplane theory to evaluate Prandtl's interference factor, which accounts for the effect of one wing upon the airflow over the other.

For delta wings, the effects of nonlinear vortex lift are important. Details of how to calculate induced drag can be found in Küchemann (1978) and McCormick (1979). Charts and formulas are available which account for sweepback angle, aspect ratio, and angle of attack.

Once the aspect ratio for the equivalent monoplane has been determined (eAR), the basic design process can continue, following the theory developed in Part 2.

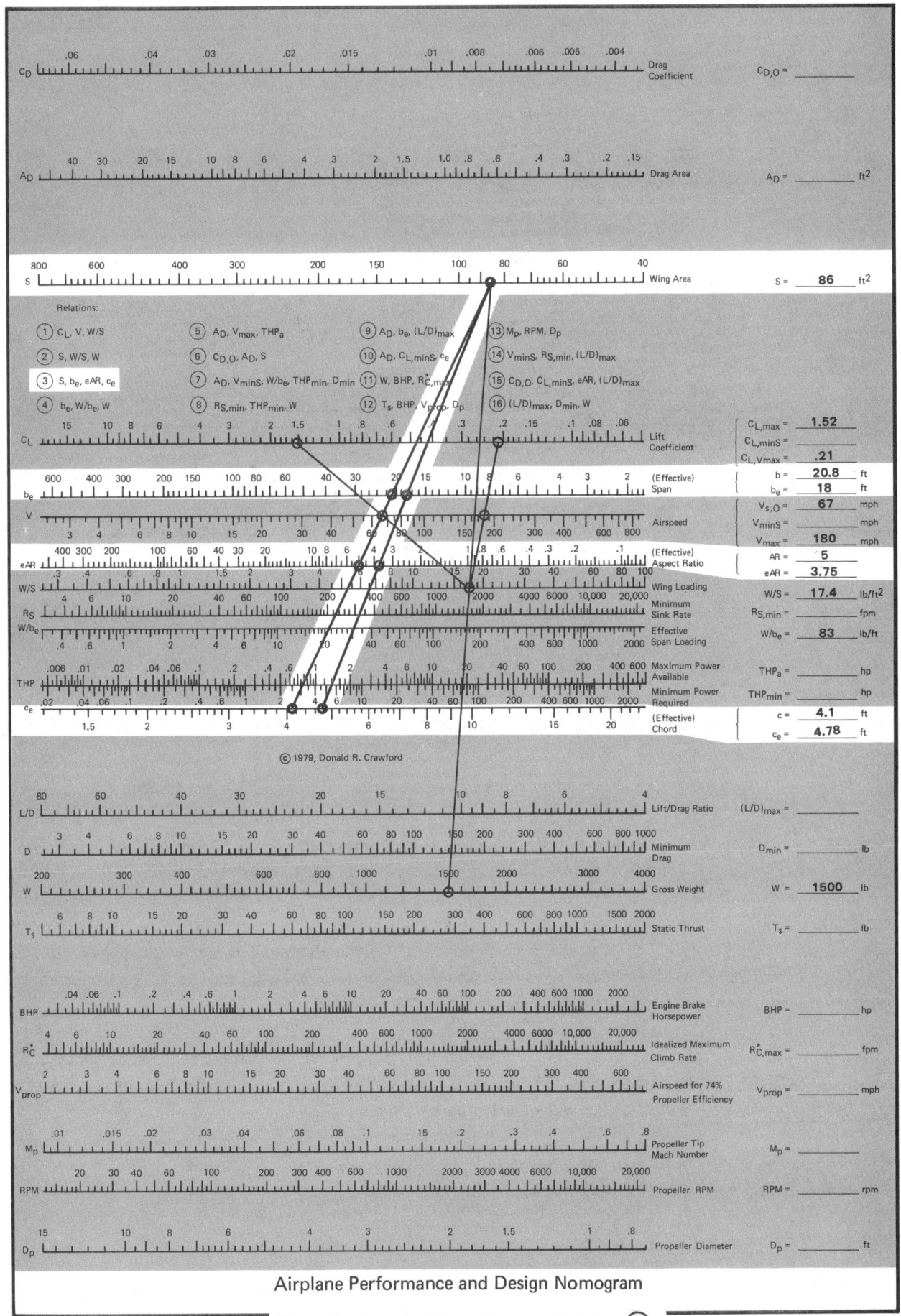

Airplane Performance and Design Nomogram

Figure 8. Sample Calculation for Relation ③.

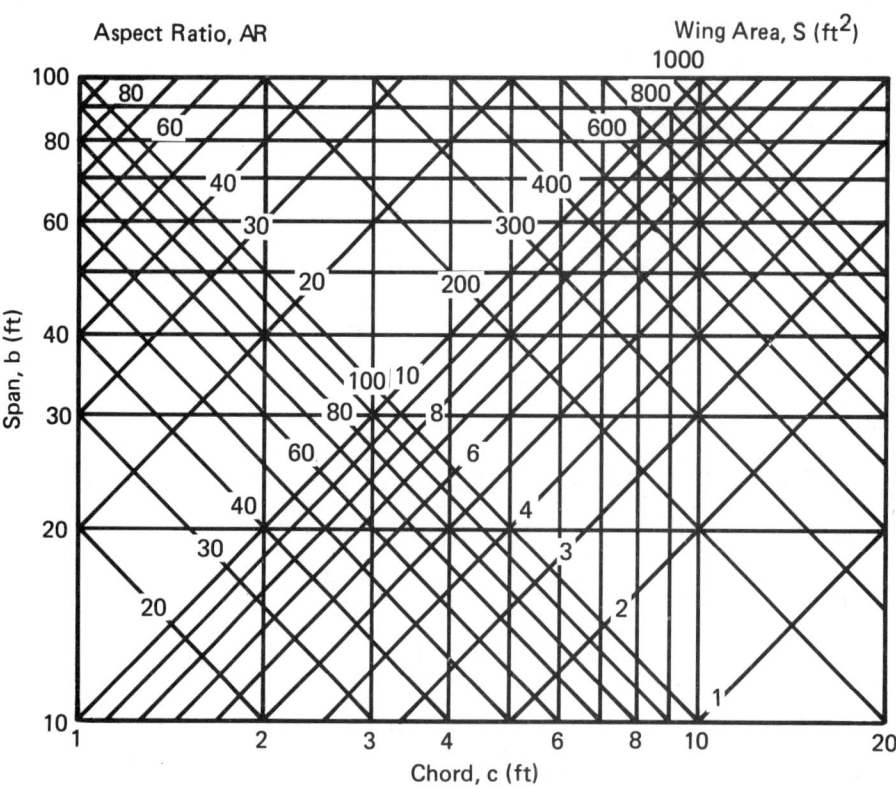

Figure 9. Relation ③: Wing Area, Span, Aspect Ratio, Chord. $S = b \cdot c$, $AR = b/c = b^2/S$.

Relation ④: Effective Span, Effective Span Loading, Gross Weight

This relation defines the effective span loading, W/b_e, in terms of the gross weight, W, and the effective span, b_e. The geometric span is first transformed into the effective span using the airplane efficiency factor and the method of relation ③. The effective span loading is an important parameter used to determine the power required for level flight. This is then used to find the sink rate, which is used for the rate of climb calculation. For the sample problem, connect the effective span (b = 18 ft) and gross weight (W = 1500 lb) of the T-18 with a straight line on Figure 10 and read the effective span loading on the W/b_e-scale (W/b_e = 83 lb/ft).

b_e, W/b_e, W	4

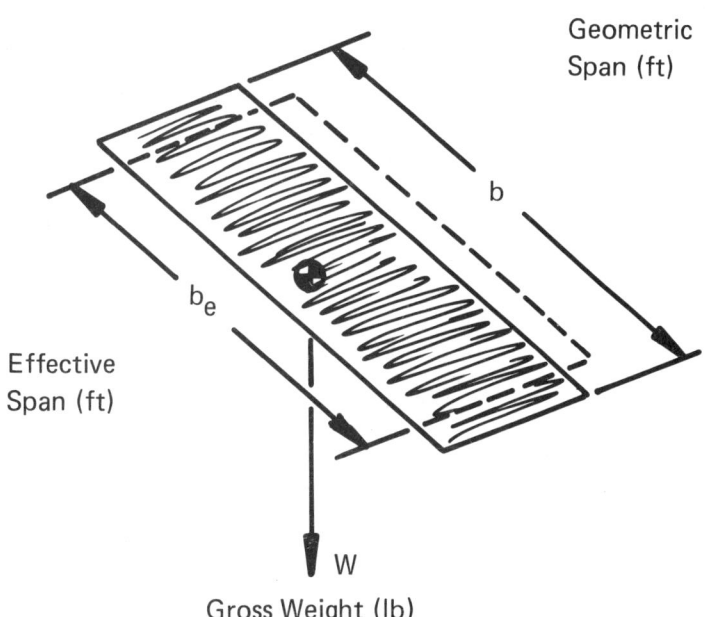

Geometric Span (ft)

b

b_e

Effective Span (ft)

W

Gross Weight (lb)

Effective Span Loading: W/b_e (lb/ft)

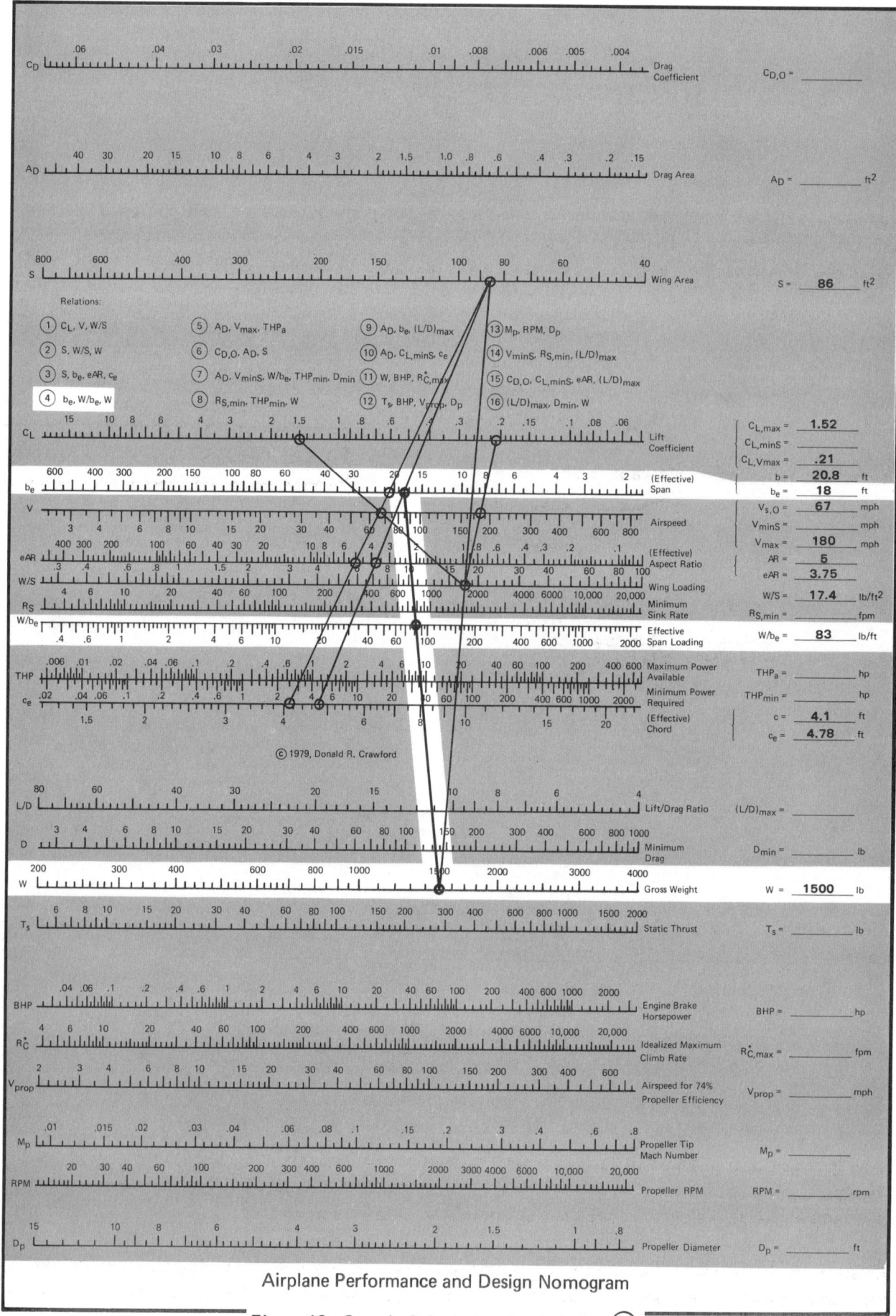

Figure 10. Sample Calculation for Relation ④.

Relation ⑤: Drag Area, Maximum Level Speed, Available Thrust Horsepower

This relation can be used to give a rough estimate of the drag area necessary to satisfy the requirement for maximum speed from a given engine-propeller combination. The drag area is defined as the area of a flat plate placed normal to the freestream airflow that will produce the same drag as the complete airplane.

Equivalent Flat Plate Drag Area.

The mathematical relationship between available thrust horsepower THP_a (hp), drag area, A_D (ft^2), and maximum level flight speed, V_{max} (mph), is

$$THP_a = \frac{A_D V_{max}^3}{146625}$$

for sea level conditions. If other altitudes are desired, the right hand side of the equation needs to be multiplied by the density ratio, $\sigma = \rho/\rho_{SL}$. However, if the sea level calculations are made first, the plastic template can be used to make rapid estimates of the effects of altitude.

Relation ⑤ is also useful for determining the size of the engine required if the drag area and the desired maximum level speed are known. Turning this around, we can ask ourselves, "How fast will it go?" if we know the powerplant size, propeller efficiency, and an estimate of the drag area.

As a first approximation we can choose a propeller efficiency of 0.80. Then, we can find how much thrust horsepower, THP_a, is available by multiplying this efficiency times the engine brake horsepower, BHP. For the sample calculation, assume that we have selected a 150-hp Lycoming engine. The thrust horsepower is 120 hp, which can then be located on the THP_a-axis of the nomogram in Figure 11. If we desire a maximum level speed of 180 mph, we can calculate the maximum allowable drag area from the nomogram. Locate the speed on the V_{max}-axis, draw a line connecting with THP_a, extending to the A_D axis, and read 3.0 ft^2 for the drag area. Therefore, if we want to go 180 mph or faster, we need to keep the drag area below three square feet. The method of estimating the drag area is discussed in Appendix G.

Relation ⑤ can also be presented in a chart having the form of Figure 12. The modified drag area, $\sigma A_D/\eta \phi$, is plotted as a function of the maximum level flight speed, V_{max}, with the engine brake horsepower, BHP, as a parameter. The density ratio, σ, and the power-altitude factor, ϕ, are both equal to unity at sea level and decrease with altitude. This will be discussed later in the parametric study of altitude effects. Also included in Figure 12 are the estimates based on the data for a number of aircraft (tabulated and discussed in Appendix L and based on **Jane's All the Worlds Aircraft 1977 - 1978**). To find the drag area of a particular airplane, multiply ($\sigma A_D/\eta \phi$) from the figure by an estimate for the propeller efficiency (use $\eta = 0.8$ for a first guess) together with values for the density ratio and power altitude factor ($\sigma = 1$ and $\phi = 1$ at sea level). For instance, for the T-18 we find $A_D/\eta = 3.3$ ft^2. If we assume that the propeller efficiency is 0.8 at sea level, then the drag area is 2.64 ft^2. From Figure 12 we see that an increase in power with the same drag area will give a higher level flight speed. Keeping the power constant, we can go faster by decreasing the drag area.

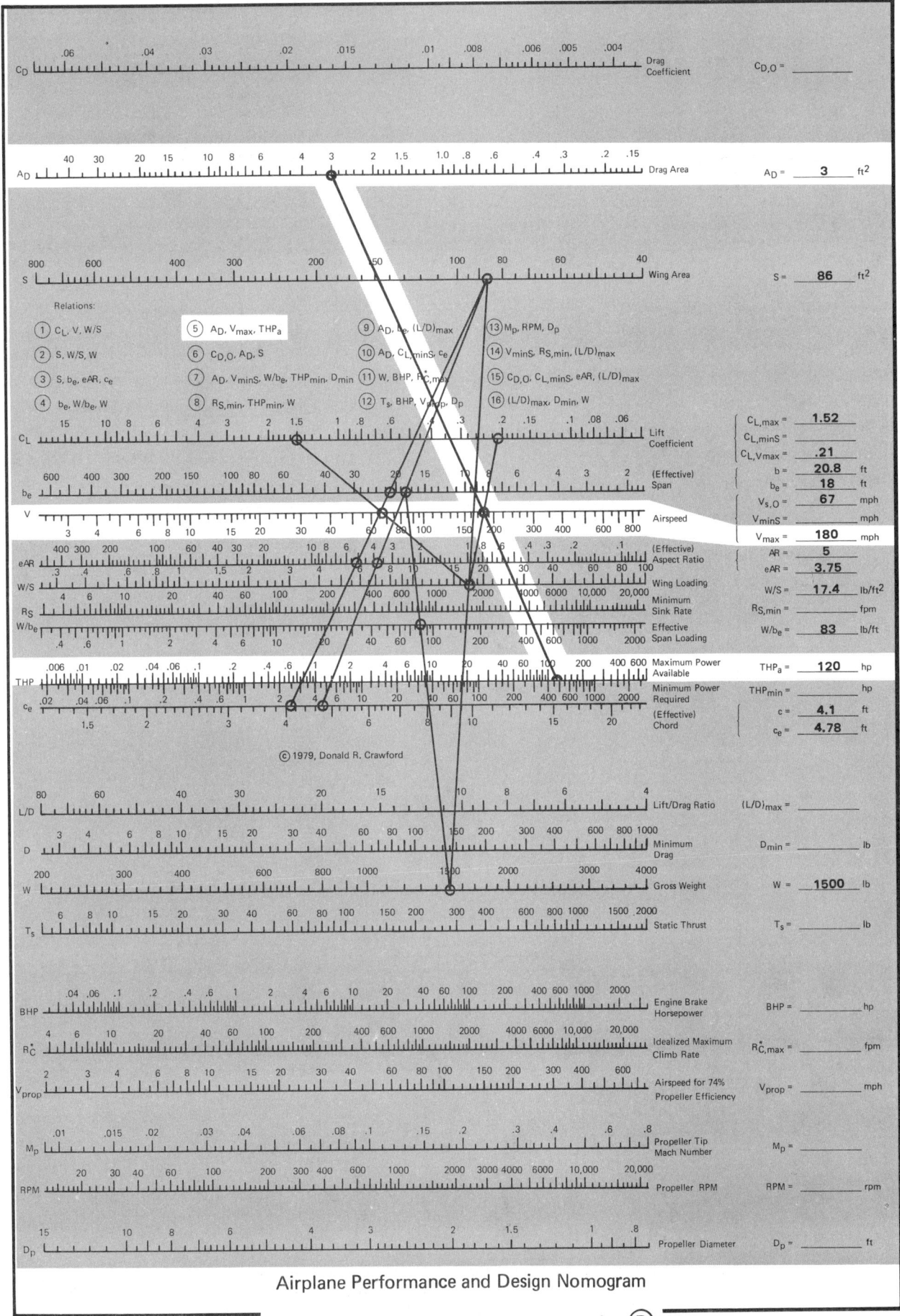

Figure 11. Sample Calculation for Relation 5.

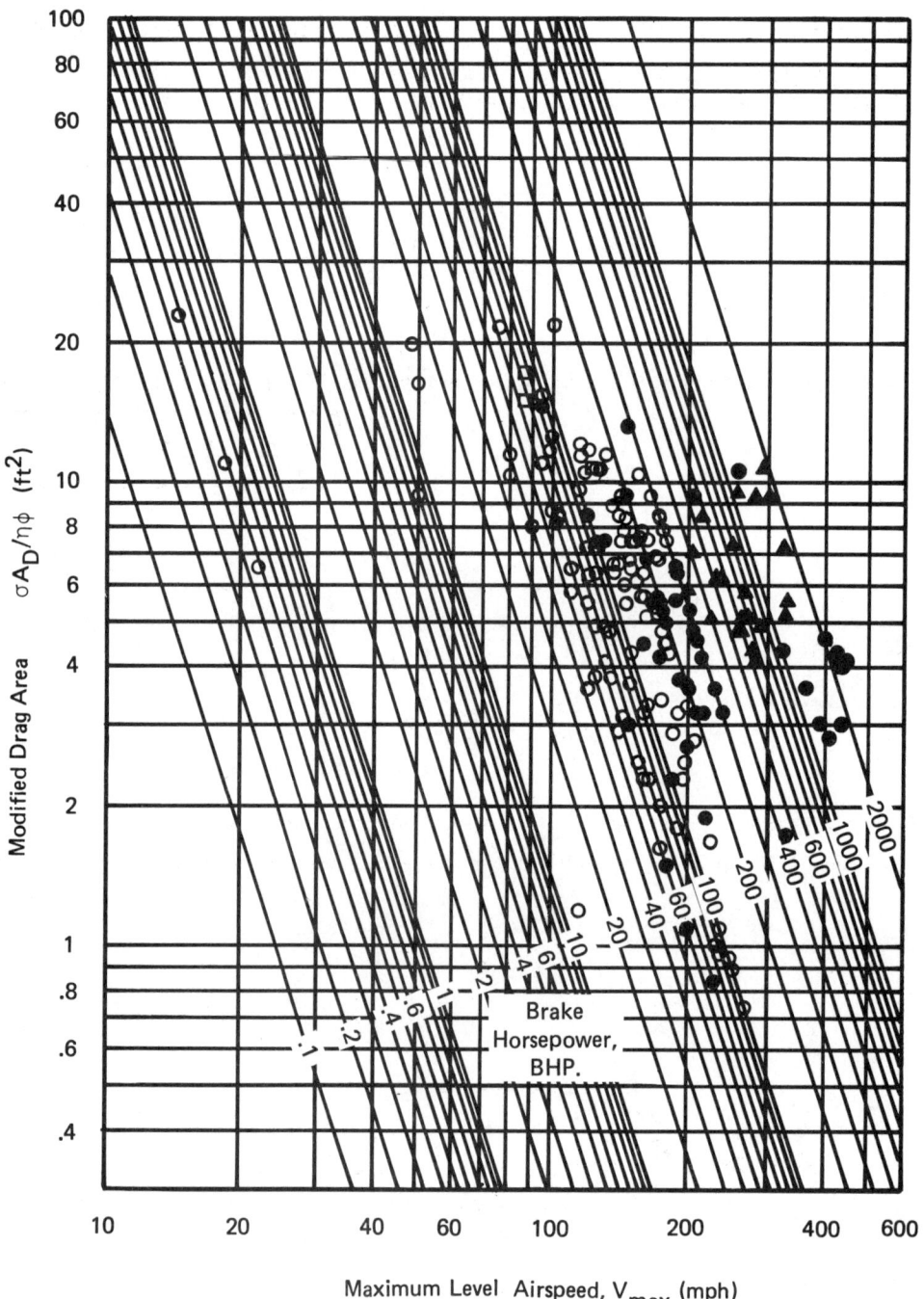

Figure 12. Relation ⑤: Drag Area, Maximum Level Airspeed, Thrust Horsepower at Sea Level $\quad THP_a \approx A_D V_{max}^3/146625$.

Relation ⑥: Zero-Lift Drag Coefficient, Drag Area, Wing Area

This relation is essentially the definition for the drag area in terms of the zero-lift drag coefficient and the wing area. $A_D = C_{D,O}S$. The drag force is expressed in terms of the density ratio, σ, drag coefficient, C_D, wing area, S, (ft^2) and relative airspeed, V (mph), by

$$D = \frac{\sigma C_D S V^2}{391} \text{ (lb)}$$

If we consider sea level conditions $\sigma = 1$ and assume that the induced drag is zero (no lift), then the zero-lift drag is

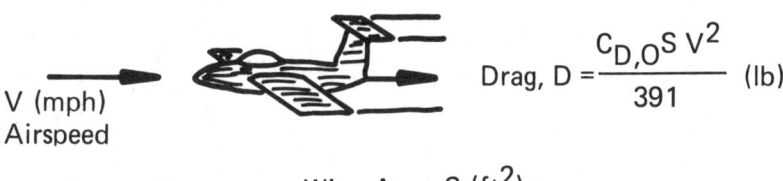

$$\text{Drag, } D = \frac{C_{D,O} S V^2}{391} \text{ (lb)}$$

For a flat plate of area A_D placed normal to the airstream, the drag coefficient is approximately equal to 1.0. Therefore, if the drag of the plate is the same as the drag of the complete airplane, we can equate A_D and $C_{D,O}S$.

$$\text{Drag, } D = \frac{A_D V^2}{391} \text{ (lb)}$$

Flat Plate Area, A_D (ft^2)

For the sample calculation, the zero-lift drag coefficient is determined from the drag area (see the drag analysis described in Appendix G) and the wing area. Draw a line on the nomogram in Figure 13 between the wing area, $S = 86$ ft^2, and the drag area, $A_D = 3$ ft^2, to find a zero-lift drag coefficient, $C_{D,O} = 0.035$, on the intersection with the C_D-axis.

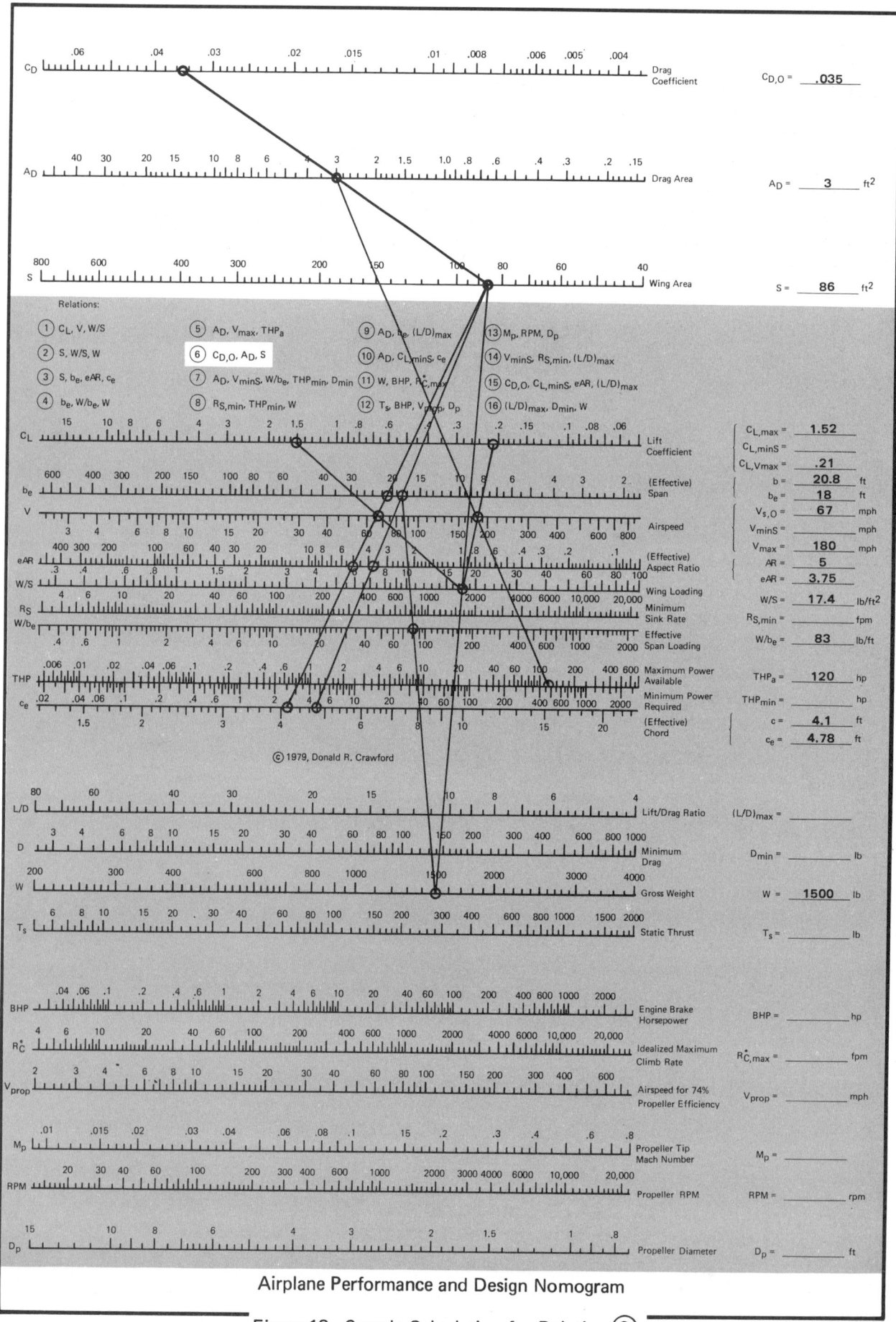

Airplane Performance and Design Nomogram

Figure 13. Sample Calculation for Relation ⑥.

Relation ⑦ :   Drag Area, Airspeed for Minimum Sink, Effective Span Loading, Minimum Power Required for Level Flight, Minimum Drag

This relation allows us to calculate more information at one time than any other relation on the Airplane Performance and Design Nomogram. Typically, knowing the drag area and the effective span loading we then use relation ⑦ to find the speed for minimum sink rate, V_{minS}, minimum power required for level flight, THP_{min}, and minimum drag, D_{min}.

For the sample calculation, the drag area is 3 ft^2 (from relation ⑤) and the effective span loading is 83 lb/ft (from relation ④). Draw a line connecting these values, intersecting with the V, THP_{min}-and D_{min}-scales. For this case, V_{minS} = 78 mph, THP_{min} = 39 hp, and D_{min} = 163 lb. These can also be calculated from the formulas tabulated at the end of the theoretical section in Table 2. The speed for minimum sink will be used to construct the curve for the sink rate as a function of the airspeed after the minimum sink rate is determined. The minimum sink rate, $R_{S,min}$, will be found with relation ⑧ using the value of the minimum power required for level flight, THP_{min}, and the gross weight, W.

If it turns out that the airspeed for minimum sink is less than the stall speed, we will have to make modifications to the shape of the curve in this region. In this case the calculated quantity, V_{minS}, is merely an idealized speed and is used only to construct the sink rate versus speed curve using the plastic template. This modification process will be explained later in the section where the use of the sink-rate template is discussed.

A_D, V_{minS}, W/b_e, THP_{min}, D_{min} — 7

Each of the equations that make up relation ⑦ can be broken down into individual charts. These are given in Figures 15, 16, and 17 for the drag area, A_D versus V_{minS}, THP_{min}, and D_{min}, with the span loading as a parameter.

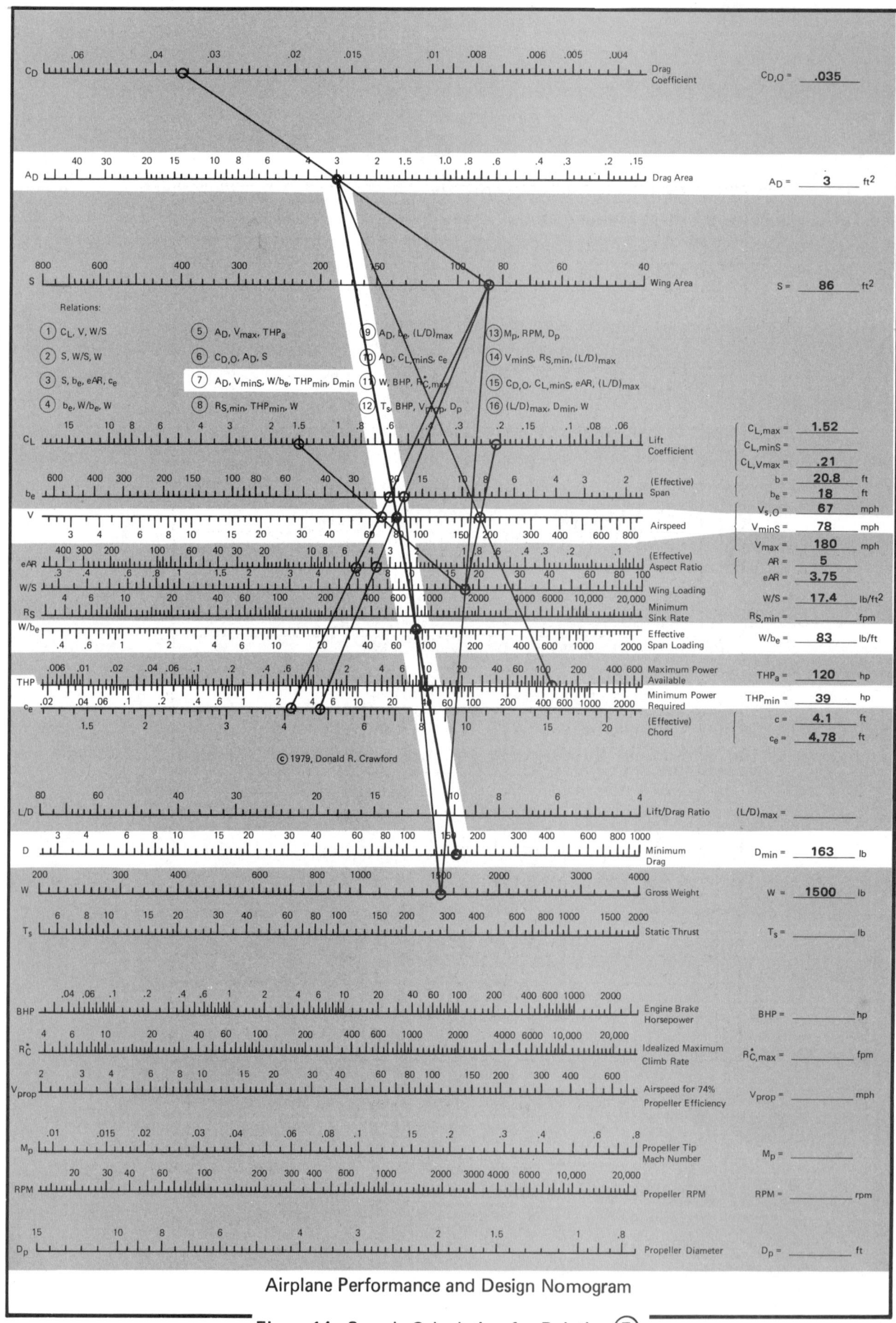

Airplane Performance and Design Nomogram

Figure 14. Sample Calculation for Relation ⑦.

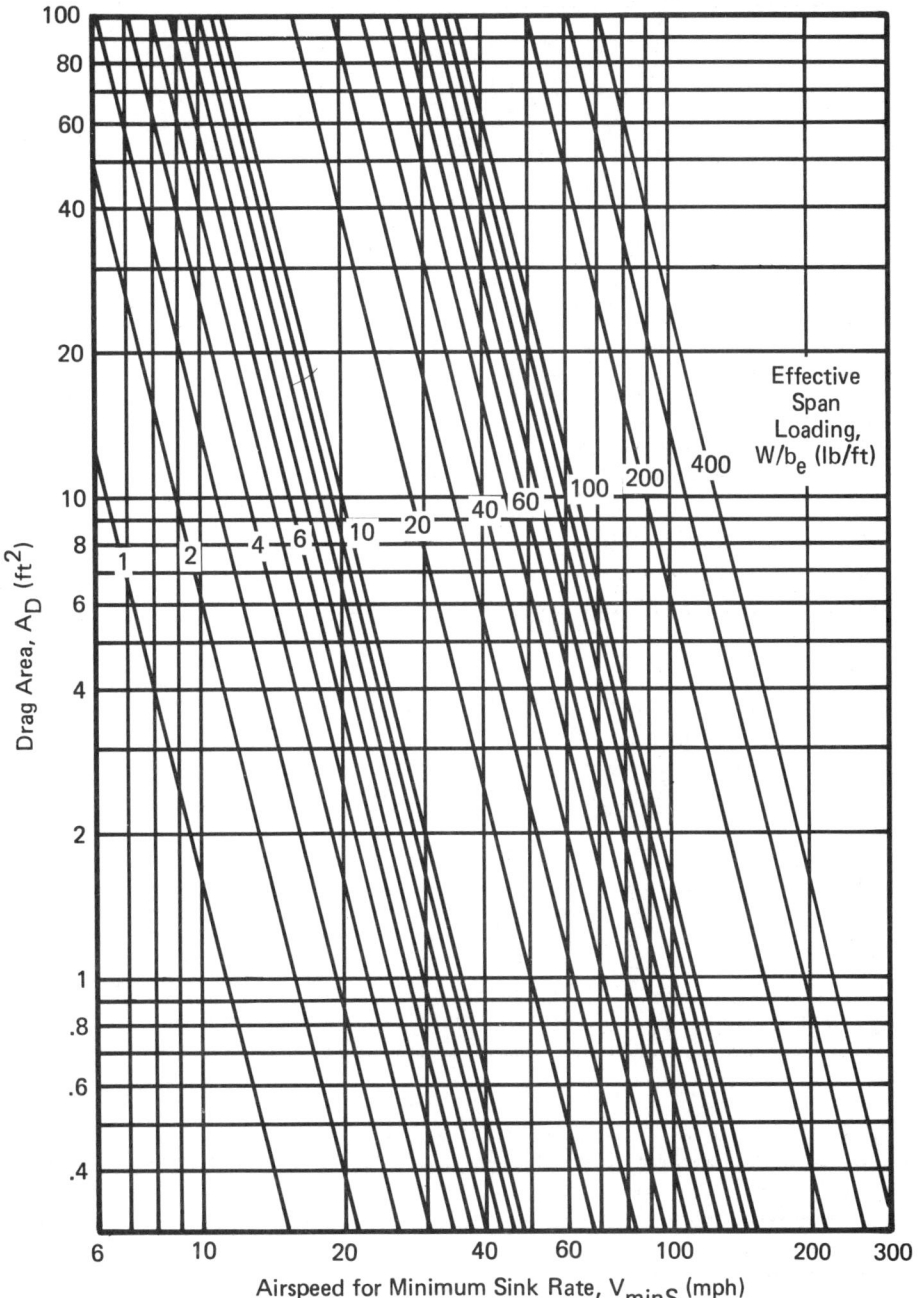

Figure 15. Relation ⑦: Drag Area, Airspeed for Minimum Sink Rate, Effective Span Loading.

$$V_{minS} = 11.29 \sqrt{W/b_e} / A_D^{1/4}$$

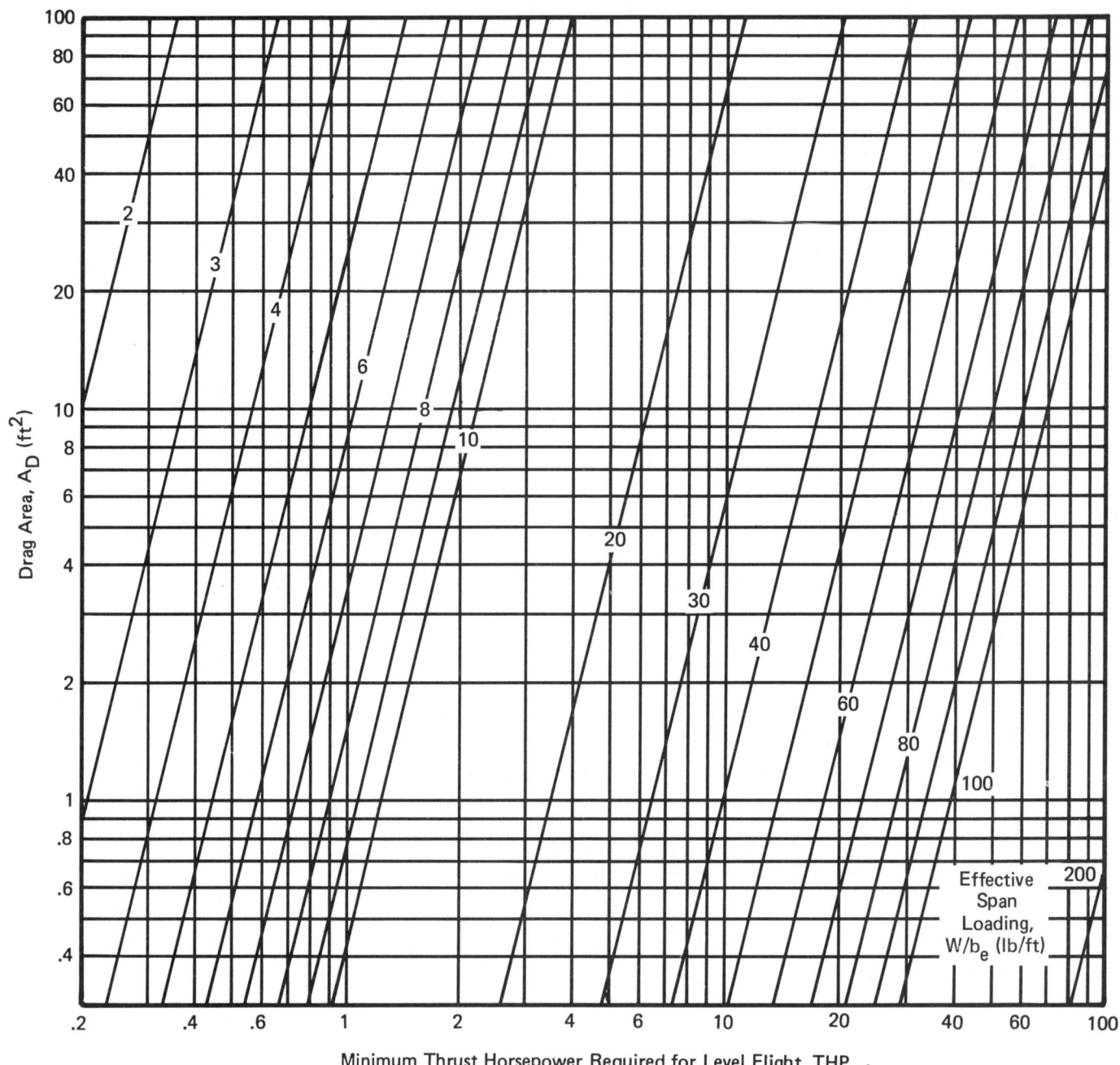

Figure 16. Relation ⑦: Drag Area, Effective Span Loading, Minimum Thrust Horsepower Required for Level Flight. $THP_{min} = 0.03921\, A_D^{1/4}\, (W/b_e)^{3/2}$.

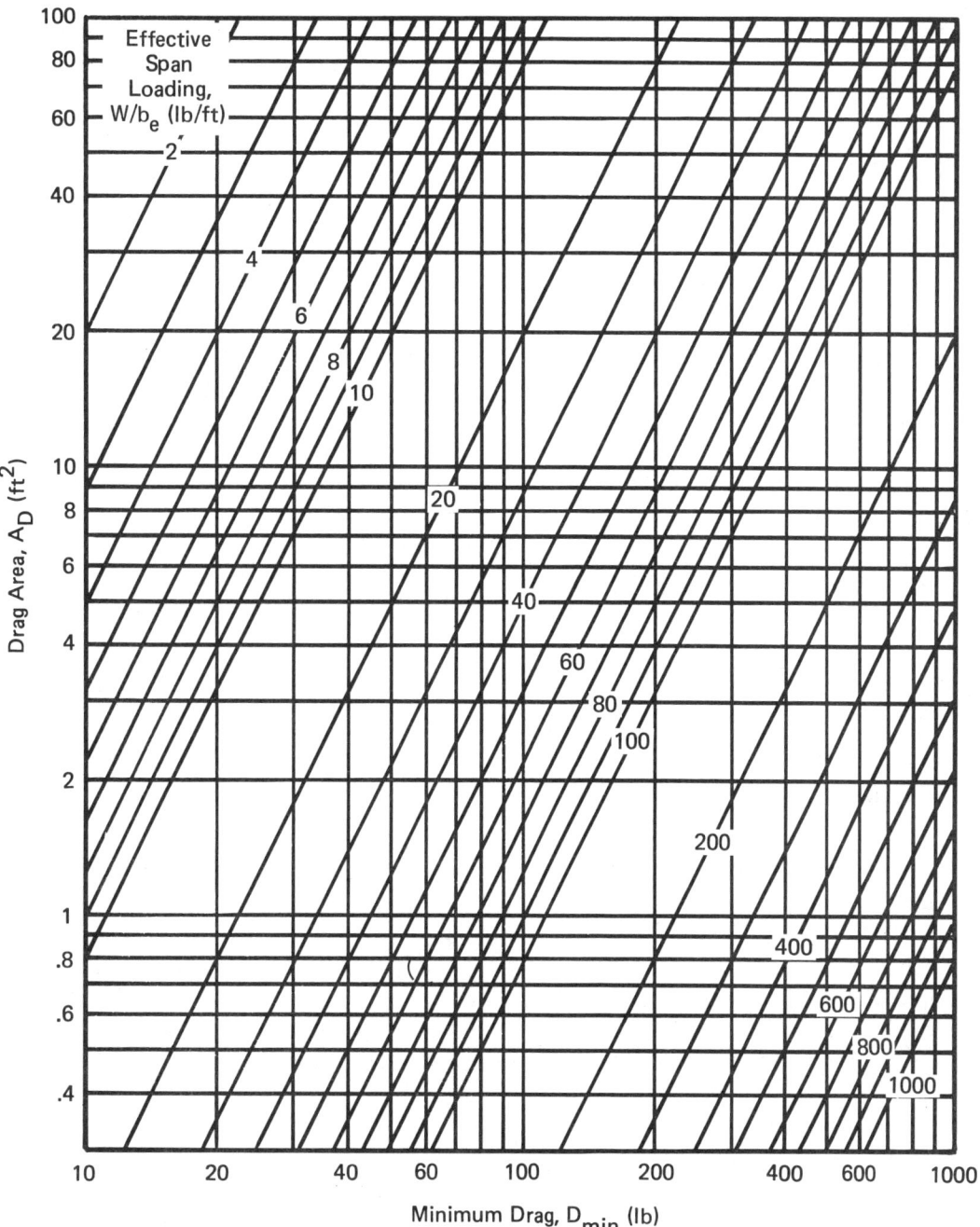

Figure 17. Relation ⑦: Drag Area, Effective Span Loading, Minimum Drag.

$$D_{min} = 1.128 \sqrt{A_D} \, (W/b_e).$$

Relation ⑧: Minimum Sink Rate, Minimum Power Required for Level Flight, Weight

If we know the minimum power required for level flight and the gross weight, we can calculate the minimum sink rate. We can also use relation ⑧ to find what minimum size engine we need for our airplane if we know the sink rate and the gross weight. The formula for the thrust horsepower is given by

$$THP = \frac{W R_S}{33,000} \text{ (hp)}$$

where the weight, W, is given in pounds, the sink rate, R_S, in feet per minute (fpm). We need to divide the thrust power by an assumed value for the propeller efficiency to estimate the engine size required from the sink-rate and weight information.

For the sample calculation, the minimum power required for level flight is 39 hp (from relation ⑦) and the gross weight is 1500 lb. Locating these on the THP_{min}- and W-scales, we draw a line on the nomogram in Figure 18 to find a sink rate of 860 fpm on the $R_{S,min}$-scale. This value will be used for the construction of the sink rate curve as a function of the airspeed. In the practical use of the Airplane Performance and Design Nomogram, values for THP_{min} and W are already available for construction of this relation since they have been found from the previous relations in the design process.

Relation ⑧ is also represented in Figure 19 where the sink rate, R_S, is plotted as a function of the power required for level flight, THP, with weight, W, as a parameter. This can also be used for relation ⑪, which has exactly the same form for the relation among BHP, $R_{C,max}^*$, and W. In fact, if we know what our sink rate is at a given speed, R_S, and we know what thrust horsepower our engine-propeller combination will deliver at that speed, THP_a, and we know the weight of our airplane, W, we can use this relation to find what the climb rate, R_C, will be:

| $R_{S,min}$, THP_{min}, W | 8 |

$$(R_C + R_S) = 33000 \frac{THP_a}{W}$$

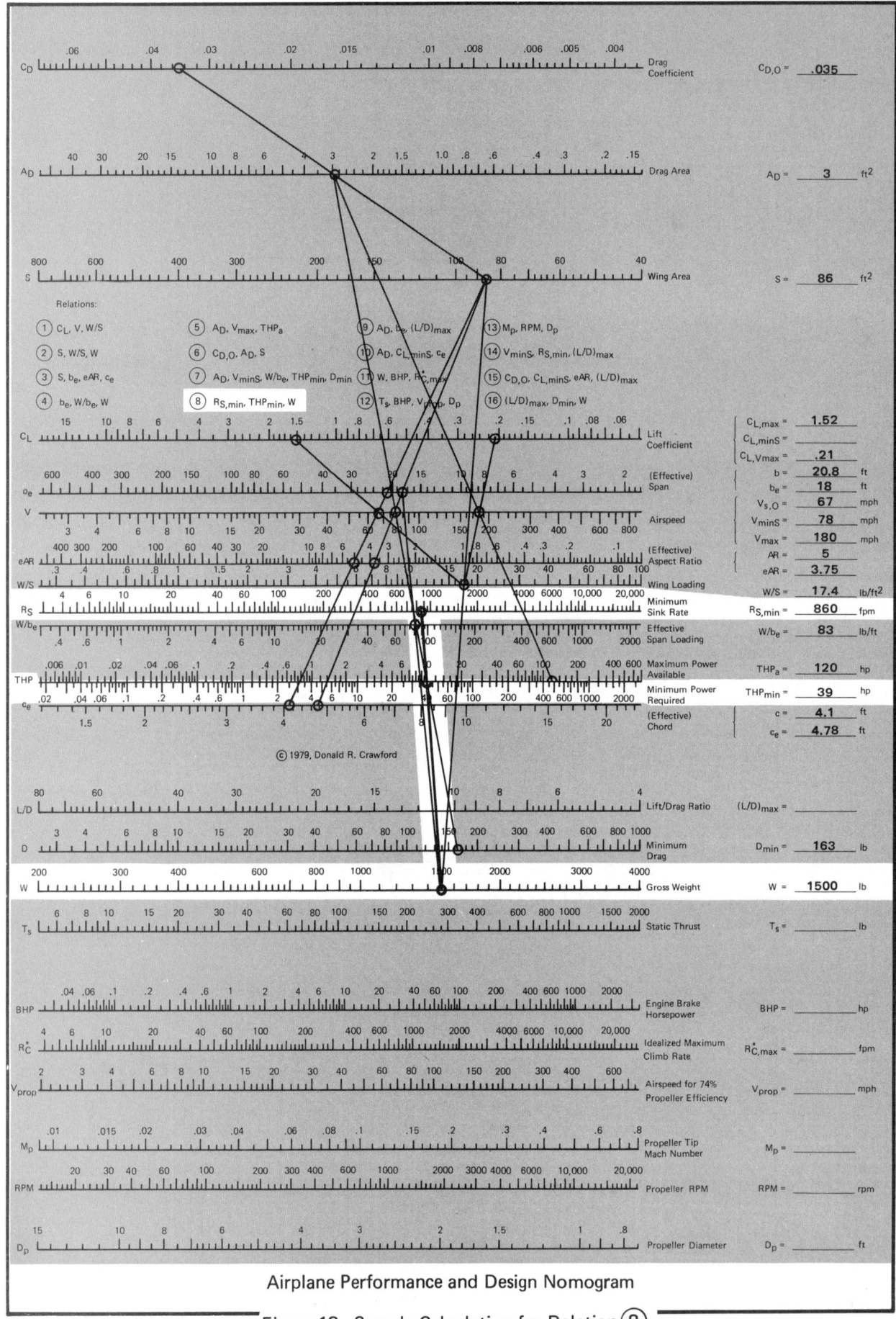

Figure 18. Sample Calculation for Relation ⑧.

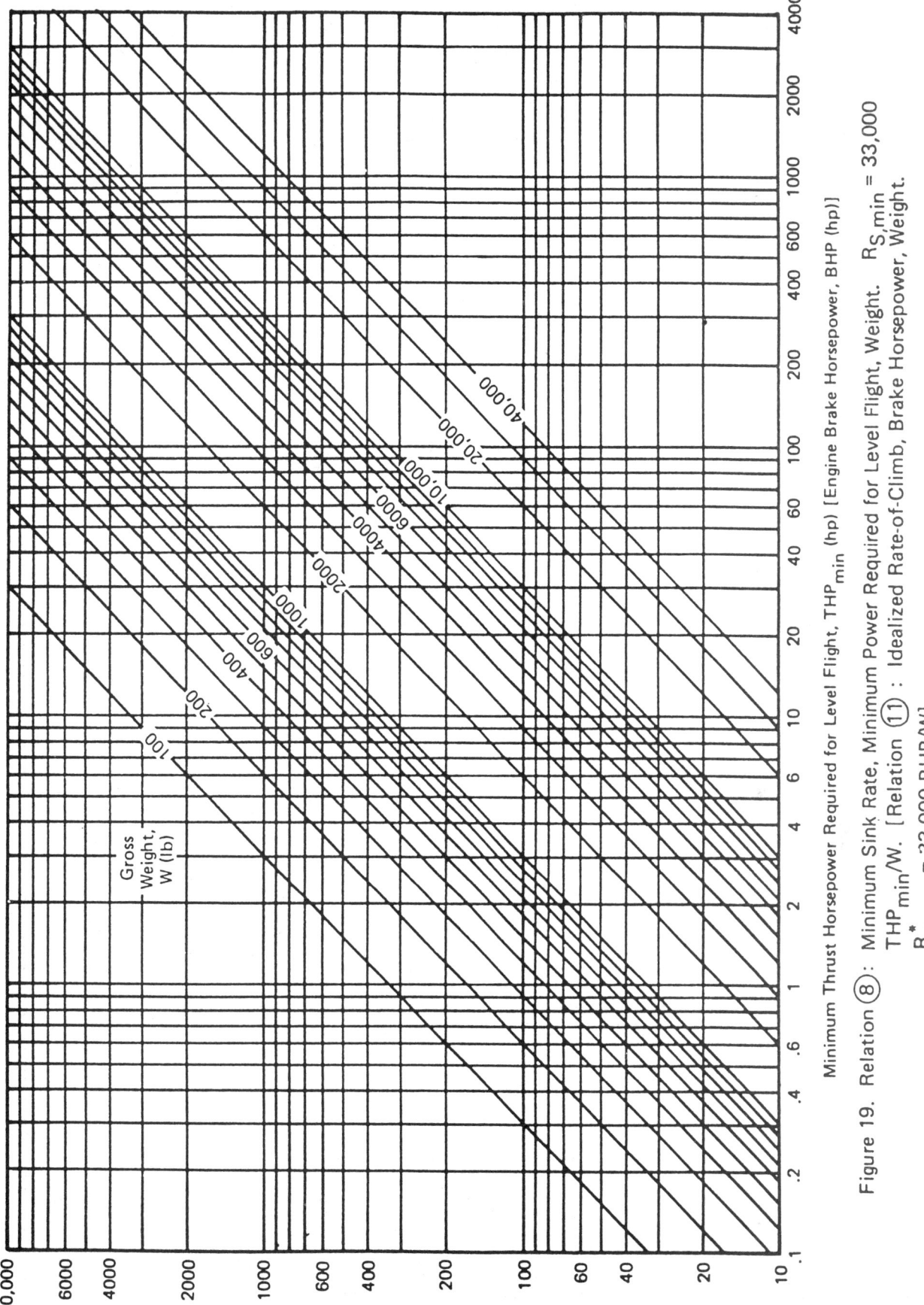

Figure 19. Relation ⑧: Minimum Sink Rate, Minimum Power Required for Level Flight, Weight. THP_{min}/W. [Relation ⑪]: Idealized Rate-of-Climb, Brake Horsepower, Weight. $R^*_{C,min} = 33,000\ BHP/W$].

Relation ⑨: Drag Area, Effective Span, Maximum Lift-to-Drag Ratio

The maximum lift-to-drag ratio is an important parameter for gliding flight. It tells us how far forward we can go for each unit of distance that we descend. If we are at an altitude of 5280 feet AGL and we have a $(L/D)_{max}$ of 10, we will be able to glide ten miles in a no-wind condition. Relation ⑨ expresses the maximum L/D ratio in terms of the effective span, b_e, and the drag area, A_D.

$$(L/D)_{max} = 0.8862 \frac{b_e}{\sqrt{A_D}}$$

given in chart form in Figure 20.

It is interesting to note that the lift-to-drag ratio does not depend on the weight, wing area or altitude. Since we have calculated the effective span to be 18 ft, from relation ③, and the drag area required for our maximum speed requirement, $A_D = 3$ ft^2 (from relation ⑤), we can connect these values on the nomogram in Figure 21 to find the maximum lift-to-drag ratio of 9.2.

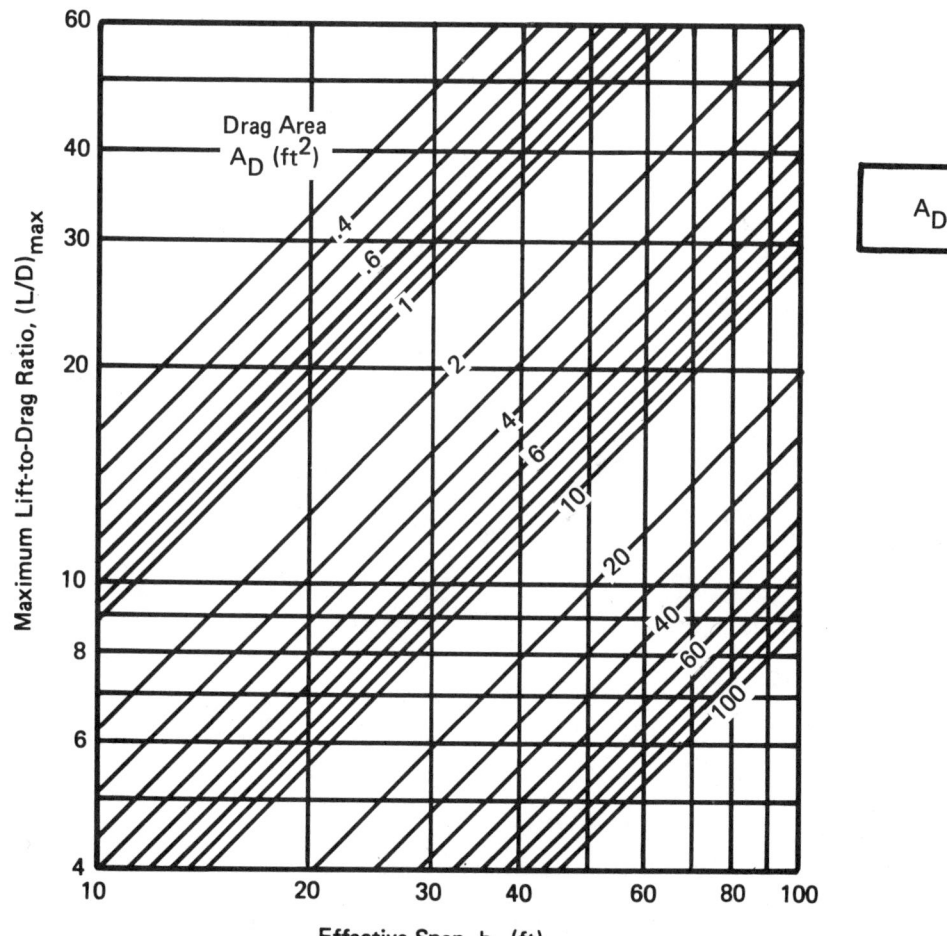

Figure 20. Relation ⑨: Drag Area, Effective Span, Maximum Lift-to-Drag Ratio.
$(L/D)_{max} = 0.8862 \, b_e/\sqrt{A_D}$

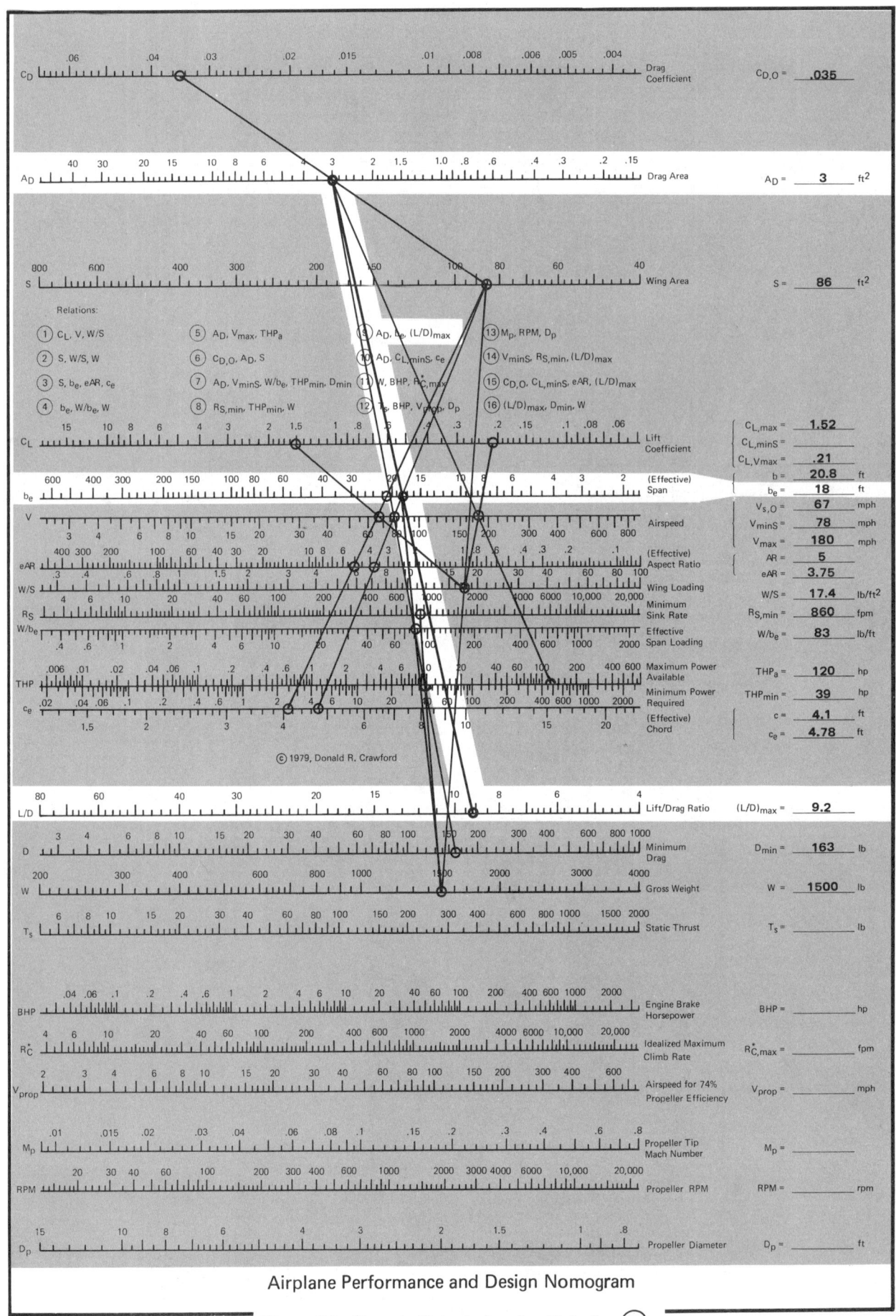

Airplane Performance and Design Nomogram

Figure 21. Sample Calculation for Relation ⑨.

Relation ⑩ : Drag Area, Lift Coefficient at Minimum Sink, Effective Chord

With this relation we can calculate the lift coefficient we expect when the airplane is flying at the speed for minimum sink. The lift coefficient, $C_{L,minS}$, is computed from the drag area, A_D, and the effective chord, c_e, which have been determined from previous relations (⑤ and ③). If we draw a line between the value of 3 for A_D and 4.75 for c_e, we find that the lift coefficient under minimum sink conditions is 1.1. This can be checked by using relation ① to connect the wing loading, W/S, and speed for minimum sink, V_{minS}, extending to the scale for the lift coefficient. If these values do not lie on a straight line, a discrepancy exists in the graphical construction that has to be corrected before the performance calculation proceeds.

If the lift coefficient under minimum sink conditions is greater than the maximum lift coefficient for the airfoil section selected, then we will have to modify the shape of the sink-rate curve in the neighborhood of the stall, as mentioned in the discussion of relation ⑦.

The formula for the lift coefficient in terms of the drag area and the effective chord is

$$C_{L,minS} = 3.07 \frac{\sqrt{A_D}}{c_e}$$

The lift coefficient under the condition for maximum L/D is smaller by the factor $1/\sqrt{3} = 0.577$. For our sample case, $C_{L,maxLD} = 0.63$, under maximum lift-to-drag conditions.

Relation ⑩ is also presented in Figure 23, where $C_{L,minS}$ is plotted as a function of the effective chord, c_e, with the drag area, A_D, as a parameter.

$A_D, C_{L,minS}, c_e$ — ⑩

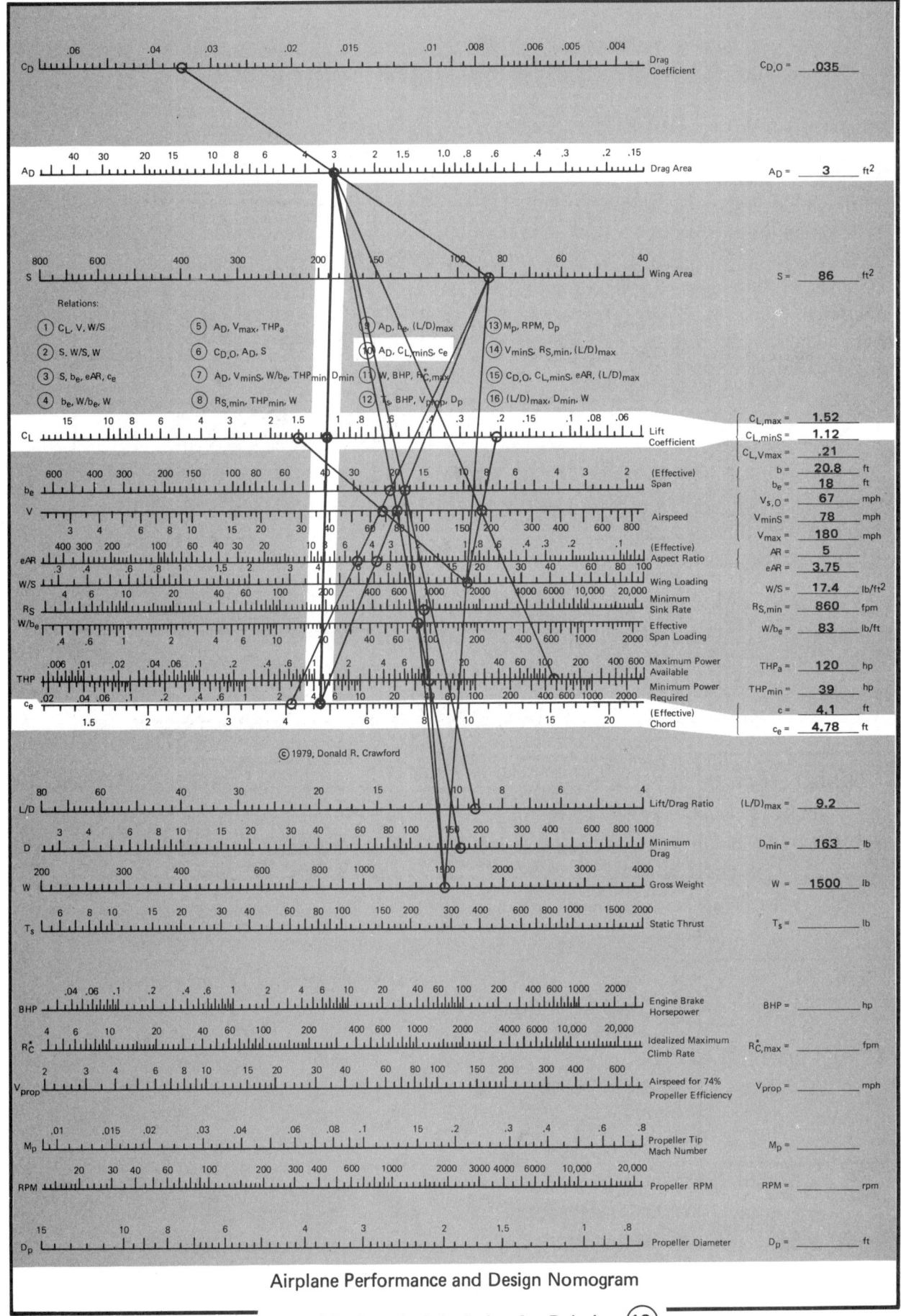

Figure 22. Sample Calculation for Relation ⑩.

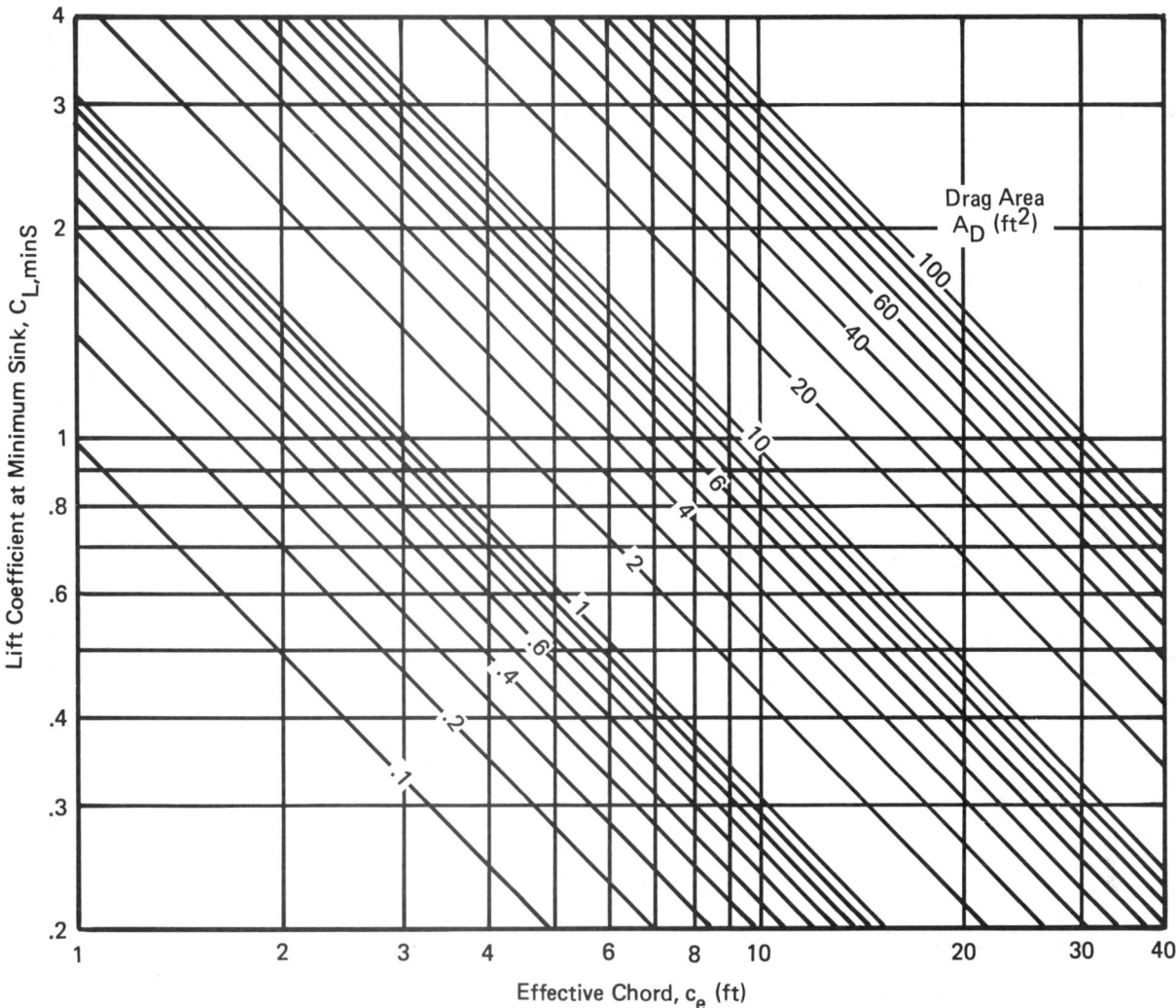

Figure 23. Relation ⑩ : Drag Area, Lift Coefficient at Minimum Sink, Effective Chord.
$$C_{L,minS} = 3.07\sqrt{A_D}/c_e$$

Relation ⑪ : Weight, Engine Brake Horsepower, Ideal Maximum Rate of Climb

The purpose of relation ⑪ is to find the idealized rate of climb, $R^*_{C,max}$, from knowledge of the engine brake horsepower, BHP, and the gross weight, W. This climb rate is the rate at which a weight equal to that of the airplane could be lifted by an engine using all of its power. This ideal climb rate will be used later with the template for the propeller efficiency to give the location of the reference point. It is used in conjunction with the reference propeller speed, V_{prop}, which is calculated with relation ⑫.

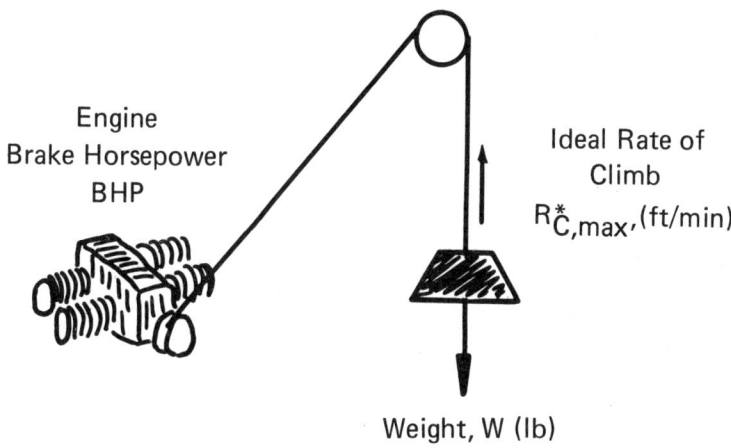

W, BHP, $R^*_{C,max}$ — ⑪

For the T-18 with a 150 hp engine, draw a straight line on the nomogram in Figure 24 connecting the BHP with the weight of 1500 lb, to find the ideal rate of climb of 3300 ft/min. The mathematical expression that is described on the nomogram is

$$BHP = \frac{R^*_{C,max} W}{33,000}$$

Figure 19 can also be used for this relation.

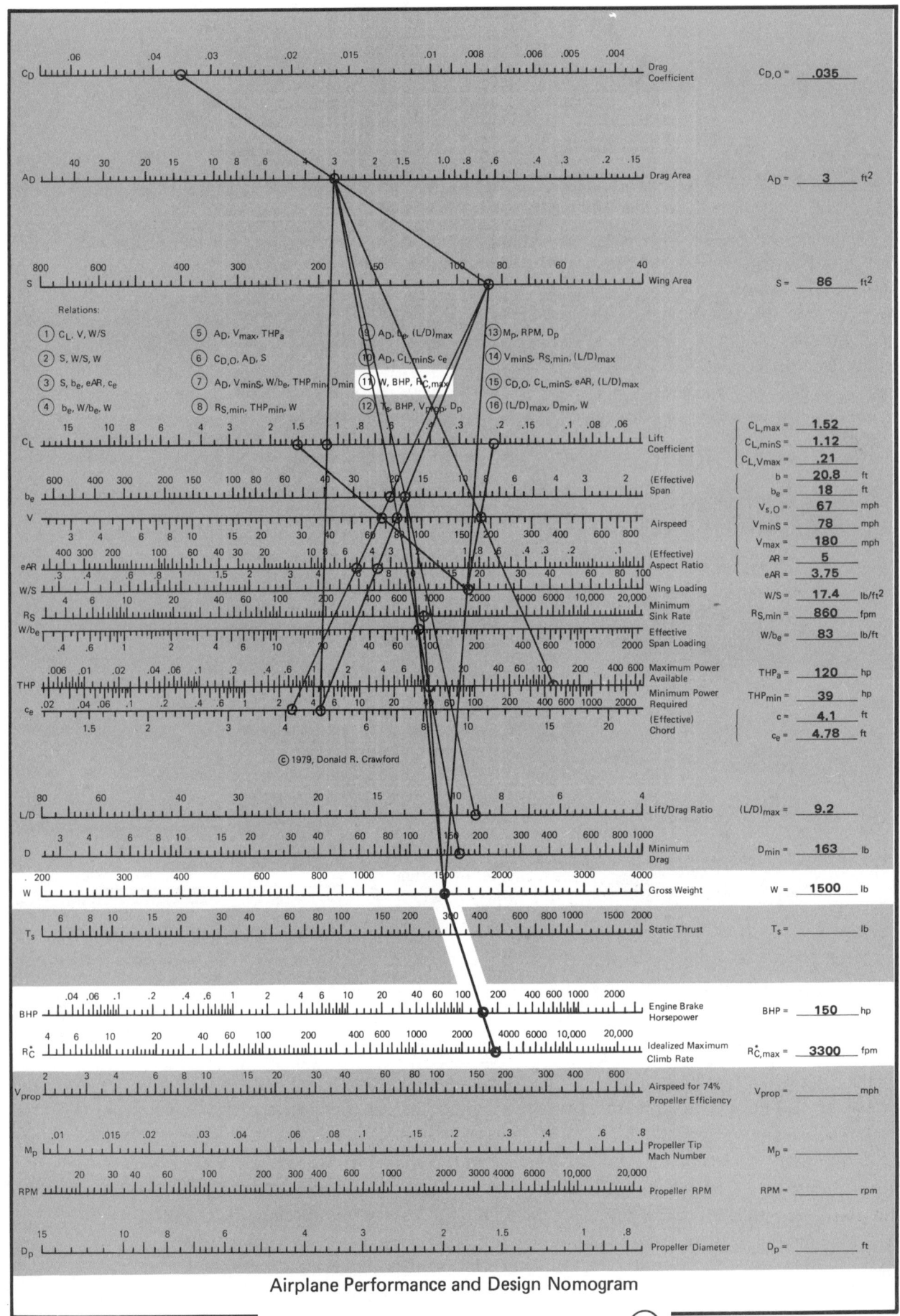

Figure 24. Sample Calculation for Relation (11).

Relation ⑫ : Static Thrust, Engine Brake Horsepower, Reference Propeller Airspeed for 74% Efficiency, Propeller Diameter

This is the key relation that comes from the idealized propeller theory of Part 2. If we have chosen an engine size, BHP, and a propeller diameter, D_p, we can find the idealized static thrust, T_s, and the reference forward speed for the ideal propeller that will result in a propeller delivering an efficiency of 74 percent, V_{prop}. This thrust and efficiency are idealizations of what can actually be expected in practice. When we use the value of V_{prop} to locate the plastic template for the propeller efficiency, we will be assuming that the propeller will deliver 85% of the idealized efficiency. This will be explained in more detail in the discussion of how to use the template. Also, the idealized static thrust cannot be fully realized, because the propeller is usually optimized for some forward speed, and therefore will be less efficient under conditions of static thrust. In fact, the propeller can have stalled blades which produce almost no thrust until the airplane starts to roll and the blades begin to bite. These limitations in the simplified theory have to be recognized as the theory is used in these calculations.

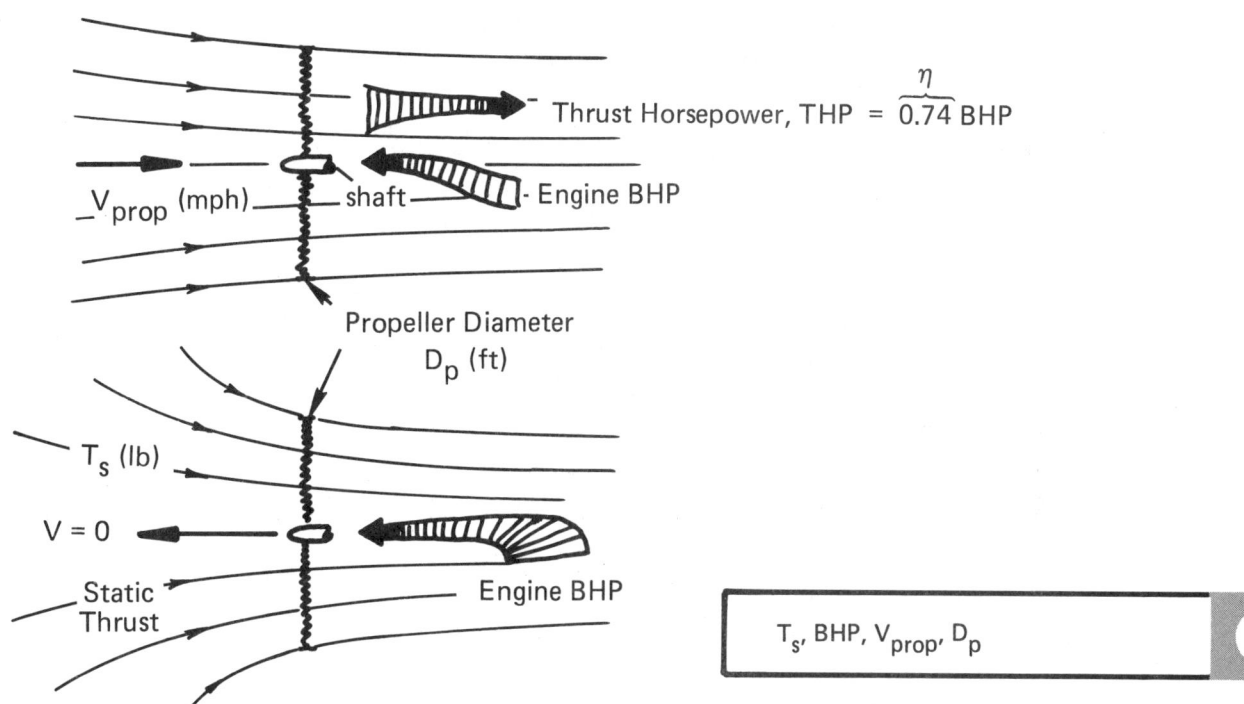

In the sample calculation, assume that the propeller is 6 feet in diameter. If we have an engine brake horsepower of 150 hp, the nomogram in Figure 25 can be used to find the idealized static thrust of 970 lb and a reference propeller airspeed, V_{prop}, of 67 mph. The calculated static thrust is an idealized value that is much larger than is obtainable. It is used when we make an analysis of the takeoff performance, using Figure 73 of Part 2 to find the thrust as a function of speed. The reference speed, V_{prop}, is used to position the plastic template to find the idealized climb rate. The difference between the idealized climb rate and the sink rate will give the rate of climb as a function of the airspeed — one of the main results of this analysis.

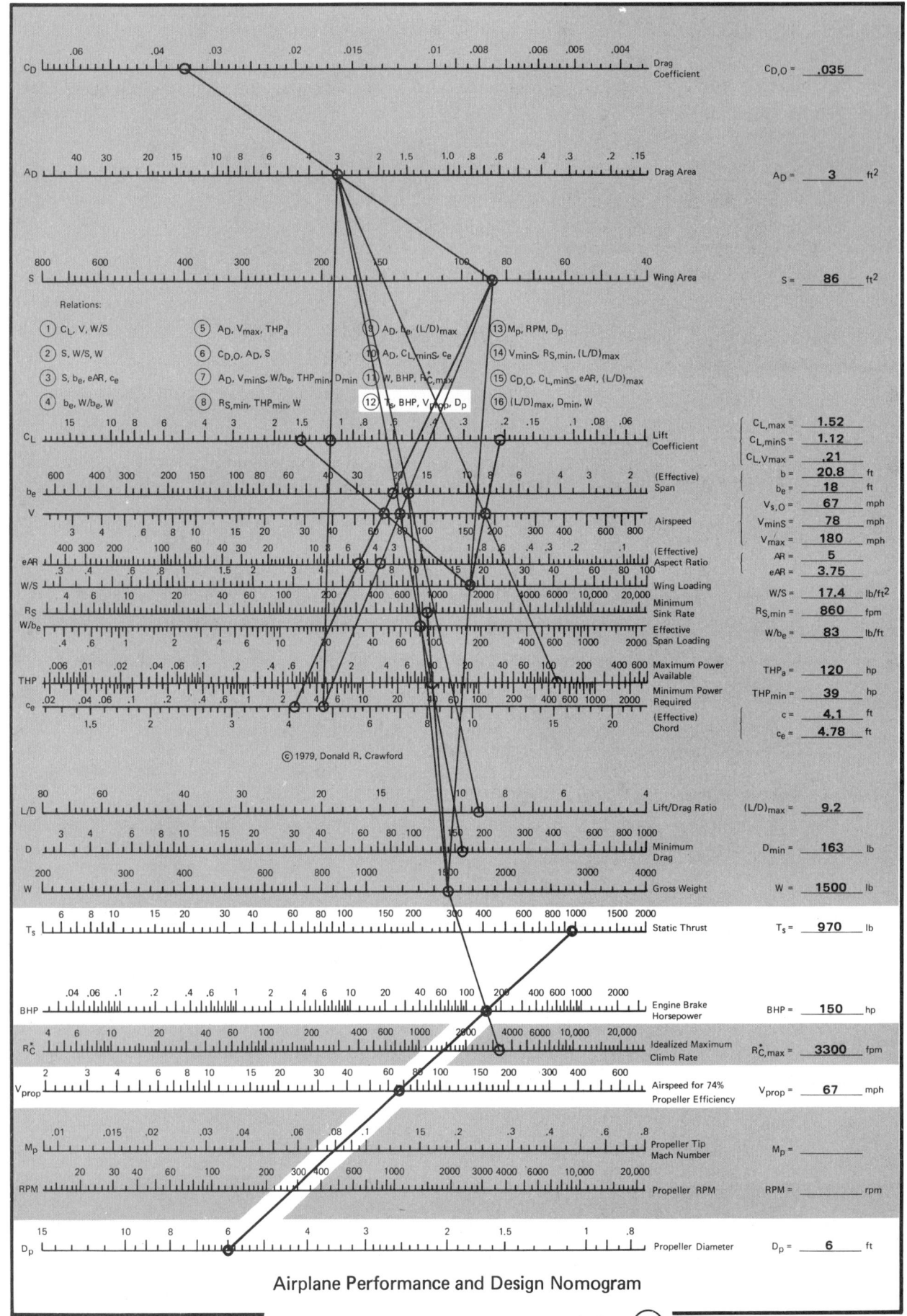

Figure 25. Sample Calculation for Relation ⑫.

The mathematical expression for the idealized static thrust at sea level is derived in Part 2 and is given by

$$T_s = 10.41\,[BHP\,D_p]^{2/3} \quad (lb)$$

where the propeller diameter, D_p, is given in feet. Similarly, the reference airspeed which corresponds to the sea-level speed for a 74% efficient idealized propeller, is given by

$$V_{prop} = 41.8 \left[\frac{BHP}{D_p^2}\right]^{1/3} \quad (mph)$$

These expressions are also individually presented in Figures 26 and 27, where we have T_s and V_{prop} plotted versus propeller diameter with BHP as a parameter.

For the best overall propeller efficiency, V_{prop} should probably be about equal to the speed for minimum sink. That way the propeller will be matched over a wide range of speeds. (For a real propeller, the efficiency will peak and then drop off again. The maximum level flight speed should be reached as the propeller efficiency peaks. This effect is not considered in the simplified analysis and has to come from more complete books on propeller theory.)

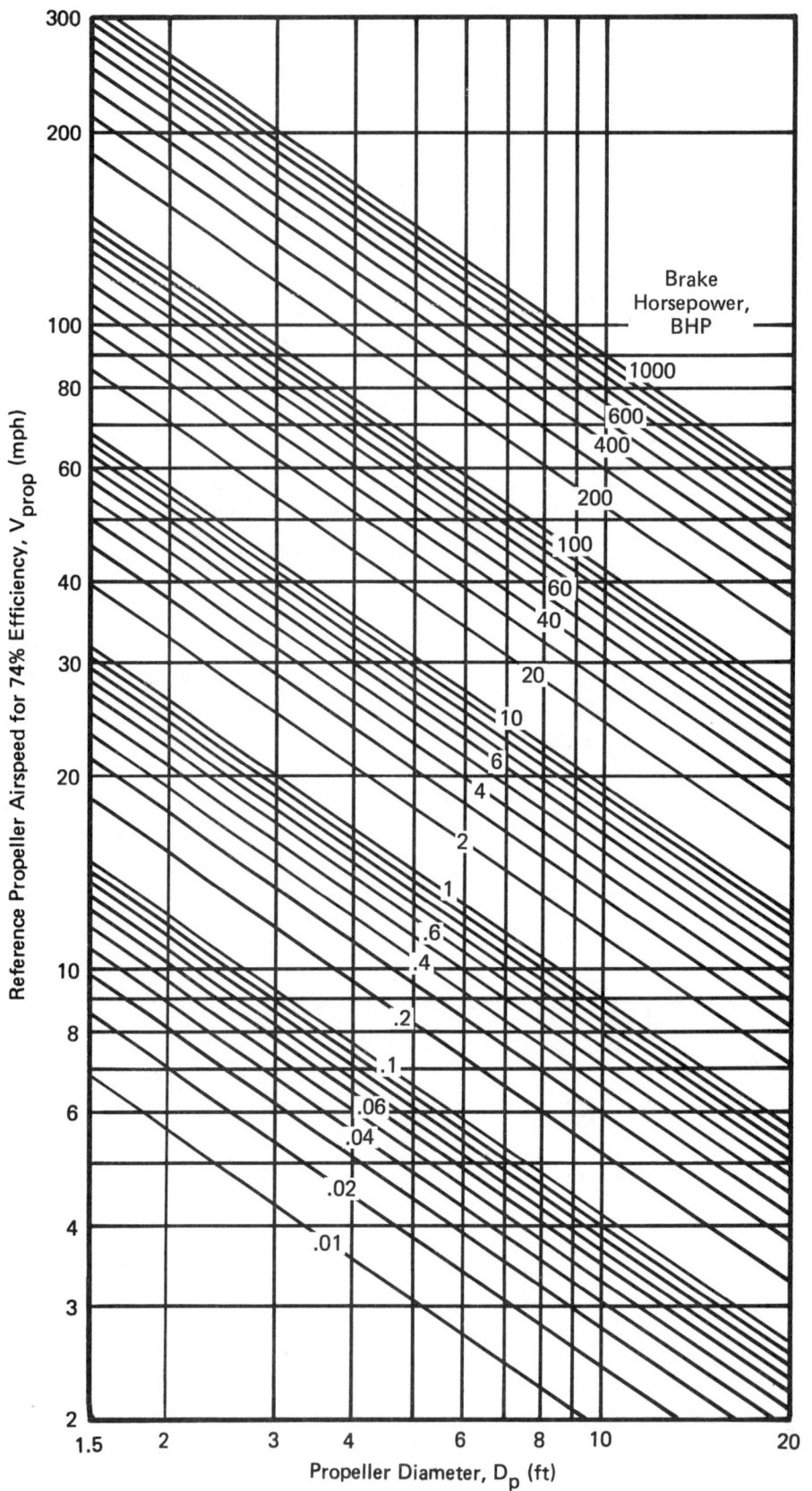

Figure 26. Relation ⑫ : Brake Horsepower, Reference Propeller Airspeed for 74% Efficiency, Propeller Diameter.

$$V_{prop} = 41.8 \, [BHP/D_p^2]^{1/3}$$

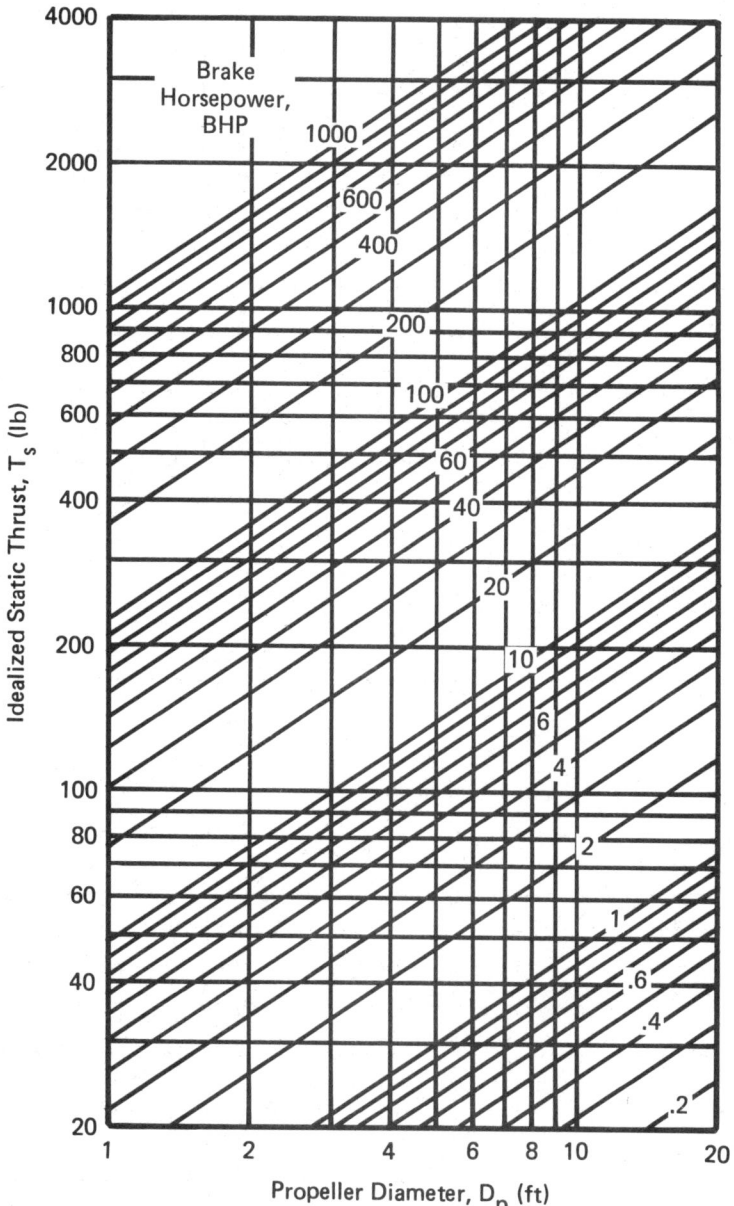

Figure 27. Relation ⑫ : Idealized Static Thrust, Engine Brake Horsepower, Propeller Diameter.

$$T_s = 10.41 \, [BHP \, D_p]^{2/3}$$

Relation ⑬ : Propeller Diameter, Propeller Rotational Speed, Propeller Tip Mach Number

This relation is used as a check to see that the propeller tip speed does not become too great. If the Mach number of the tip of the propeller exceeds 0.8, the efficiency decreases dramatically and the noise from the blades increases. If we take the propeller diameter, D_p, and the propeller rotational speed, RPM, we can find the tip speed, V_{tip}, and compare this to the speed of sound. The ratio V_{tip}/a is the propeller tip Mach number, M_p, and is given in equation form by

$$M_p = \frac{RPM\ D_p}{21,008}$$

where the propeller diameter is given in feet.

For the sample calculation, assume that the propeller rotational speed is 2700 rpm, and the propeller diameter is 6 ft. From the nomogram in Figure 28, we find that the propeller tip Mach number is 0.77, which is below the value where the tip losses start to occur.

Relation ⑬ is also represented in Figure 29, where we have plotted RPM versus propeller diameter with tip Mach number as a parameter.

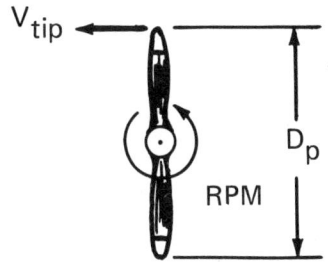

Speed of sound ≈ 1100 ft/sec.

D_p, RPM, M_p ⑬

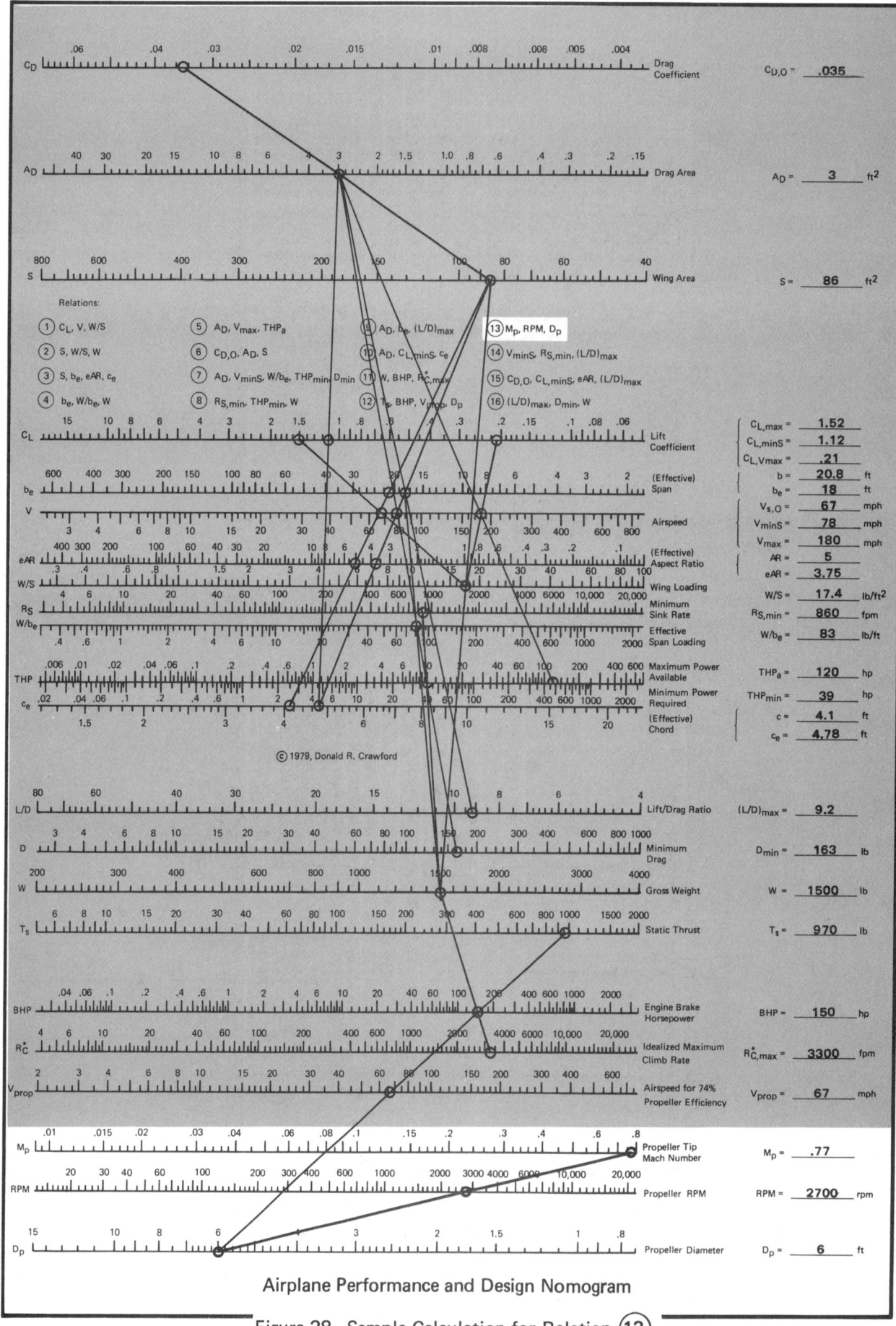

Figure 28. Sample Calculation for Relation ⑬.

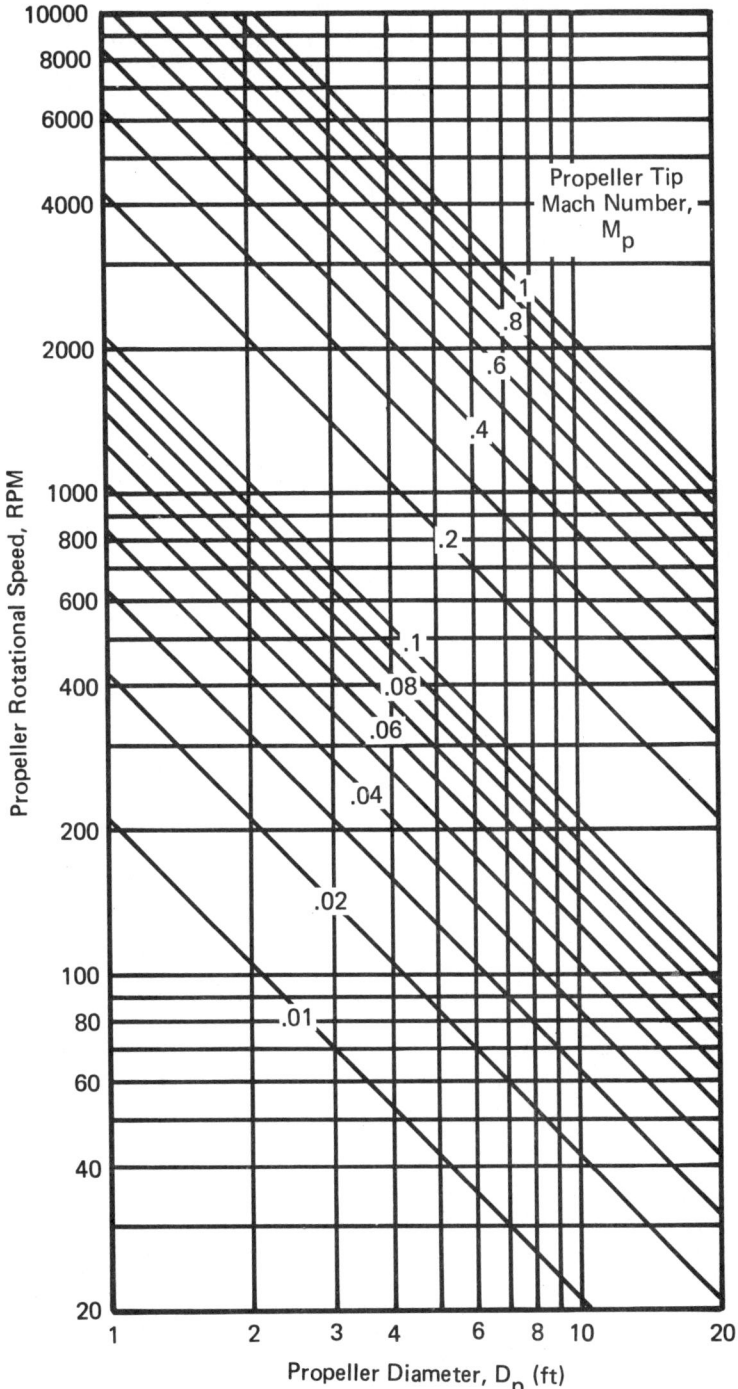

Figure 29. Relation ⑬ : Diameter, Rotational Speed, Tip Mach Number for Propellers.

$M_p = RPM\ D_p/21{,}008.$

Relation ⑭ : Airspeed for Minimum Sink Rate, Minimum Sink Rate, Maximum Lift-to-Drag Ratio

This relation is used as a consistency check of the graphical construction on the Airplane Performance and Design Nomogram. Since V_{minS} has already been calculated from relation ⑦, $R_{S,min}$ from ⑧ and $(L/D)_{max}$ from ⑨, we can connect these values to verify that they lie on a straight line. If they cannot be connected, an error exists in the construction process which must be corrected before we can have confidence in the results.

The term 88 $V_{minS}/R_{S,min}$ is the lift-to-drag ratio under minimum sink conditions. Since the lift-to-drag ratio under minimum sink conditions is related to the maximum lift-to-drag ratio, $(L/D)_{max}$, by the expression $(L/D)_{minS} = 0.866 (L/D)_{max}$, we have

$$(L/D)_{max} = 101.6 \frac{V_{minS}}{R_{S,min}},$$

where V_{minS} is the airspeed for minimum sink (mph) and $R_{S,min}$ is the minimum sink rate (fpm). Relation ⑭ is represented by the nomogram in Figure 30 as well as the chart in Figure 31.

For the sample calculation, $V_{minS} = 78$ mph, $R_{S,min} = 860$ fpm and $(L/D)_{max} = 9.2$. From the nomogram in Figure 30, we see that these results are consistent.

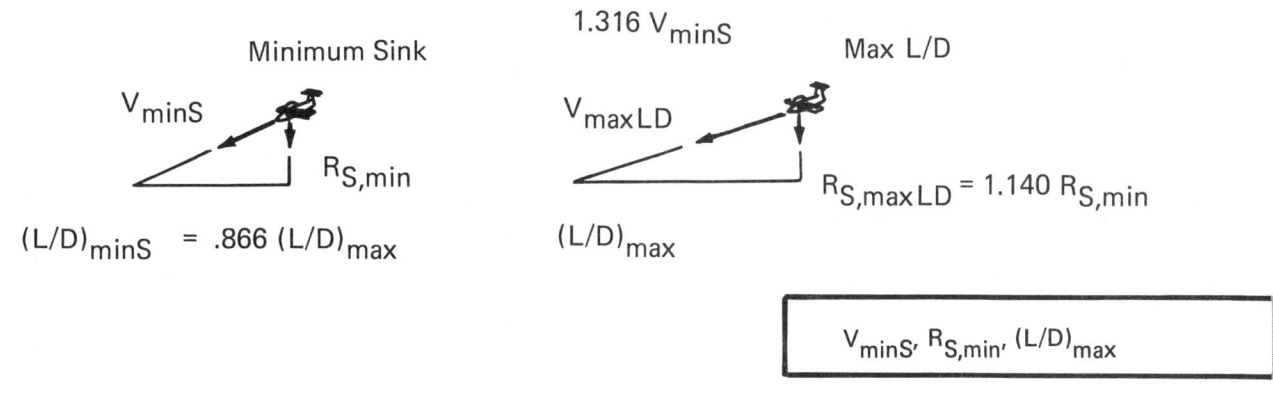

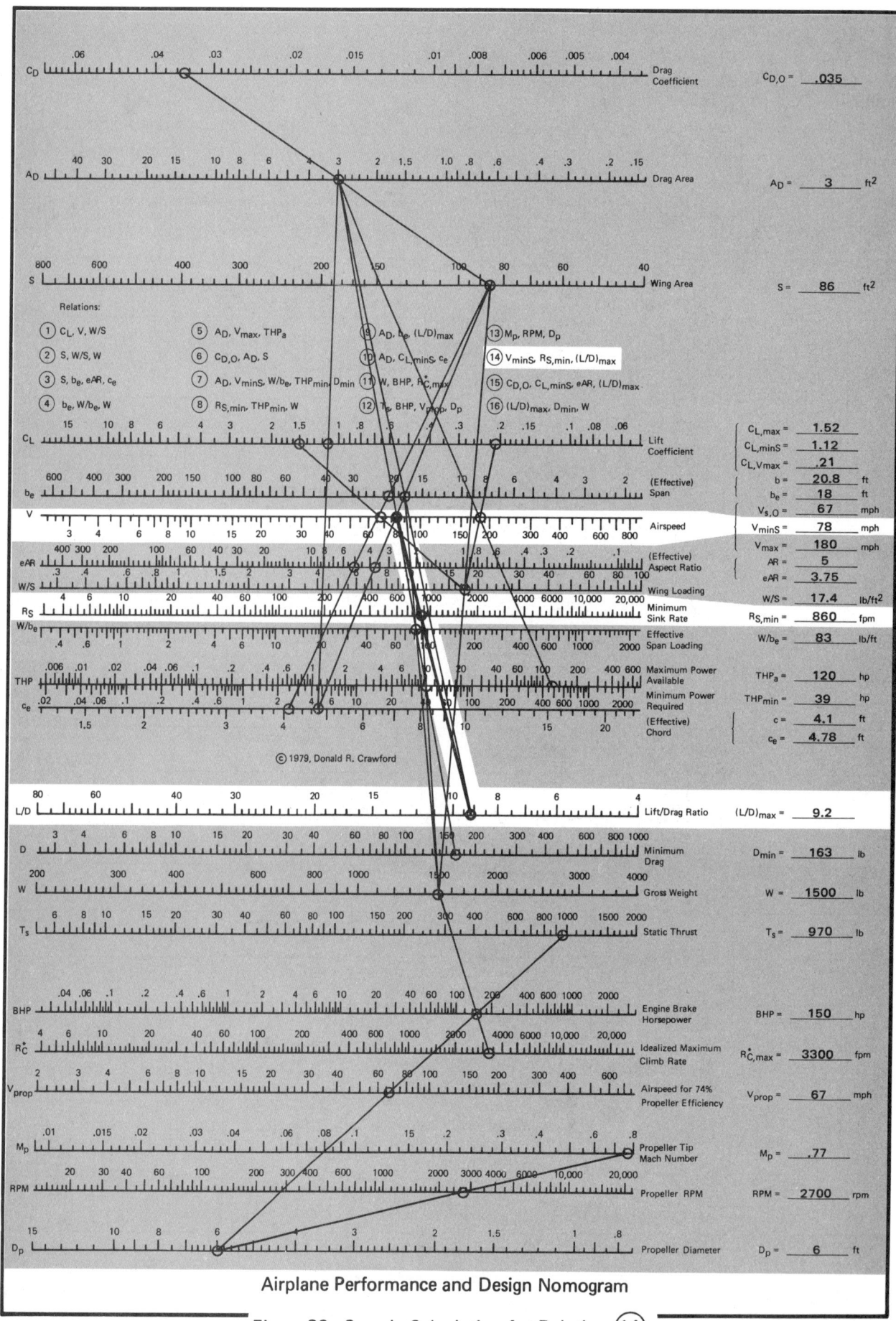

Figure 30. Sample Calculation for Relation (14).

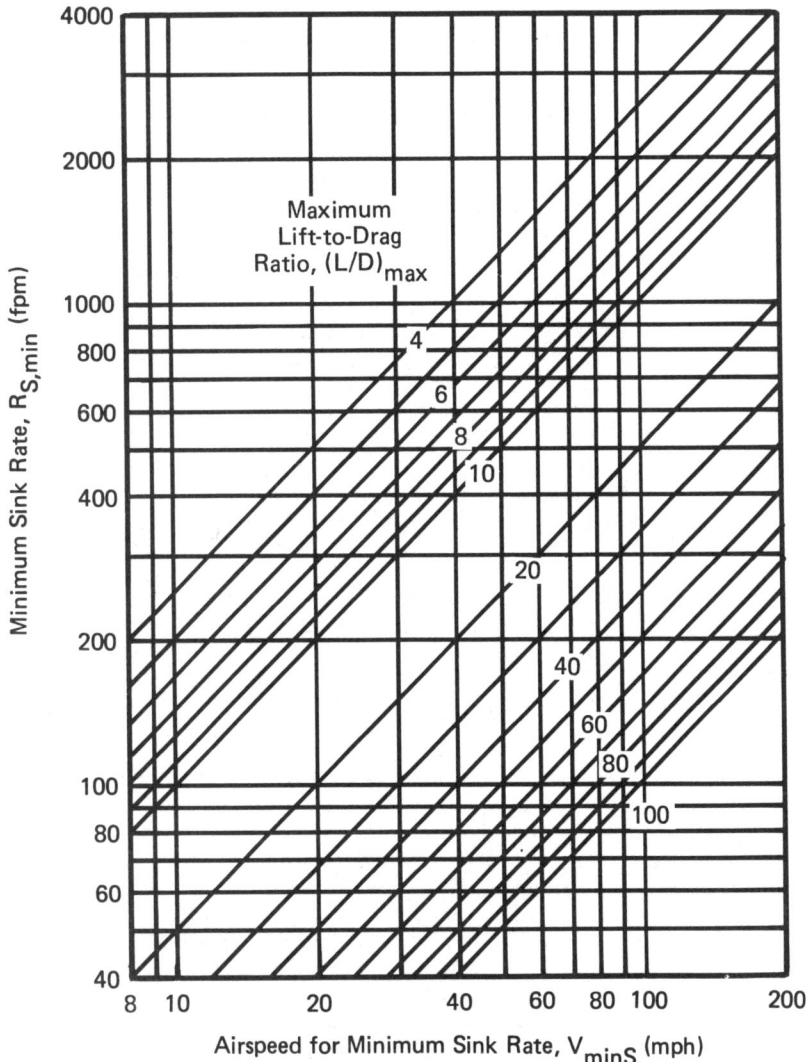

Figure 31. Relation (14) : Airspeed for Minimum Sink Rate, Minimum Sink Rate, Maximum Lift-to-Drag Ratio.

$(L/D)_{max} = 101.6\, V_{minS}/R_{S,min}$

Relation ⑮ : Zero-Lift Drag Coefficient, Lift Coefficient at Minimum Sink, Effective Aspect Ratio, Maximum Lift-to-Drag Ratio

Relation ⑮ is another cross check for the graphical construction on the Airplane Performance and Design Nomogram. We have already calculated $C_{D,O}$ from relation ⑥, $C_{L,minS}$ from ⑩, eAR from ③ and $(L/D)_{max}$ from ⑨. For a consistent set of calculations, these four quantities should lie on the same straight line on the nomogram in Figure 32. For the sample calculation for the T-18, we calculated $C_{D,O}$ = 0.035, $C_{L,minS}$ = 1.1, eAR = 3.7 and $(L/D)_{max}$ = 9.2. We can see from Figure 32 that our cross check works. If it did not, we would have to double check our graphical calculations and find the error before we continue.

The mathematical equations represented by this relation have been derived in Part 2. The lift coefficient at minimum sink is given by

$$C_{L,minS} = 3.07 \sqrt{eAR \; C_{D,O}}$$

and the equation for the maximum lift-to-drag ratio is given by

$$(L/D)_{max} = 0.886 \sqrt{\frac{eAR}{C_{D,O}}}$$

These two expressions are also presented for convenience in Figures 33 and 34.

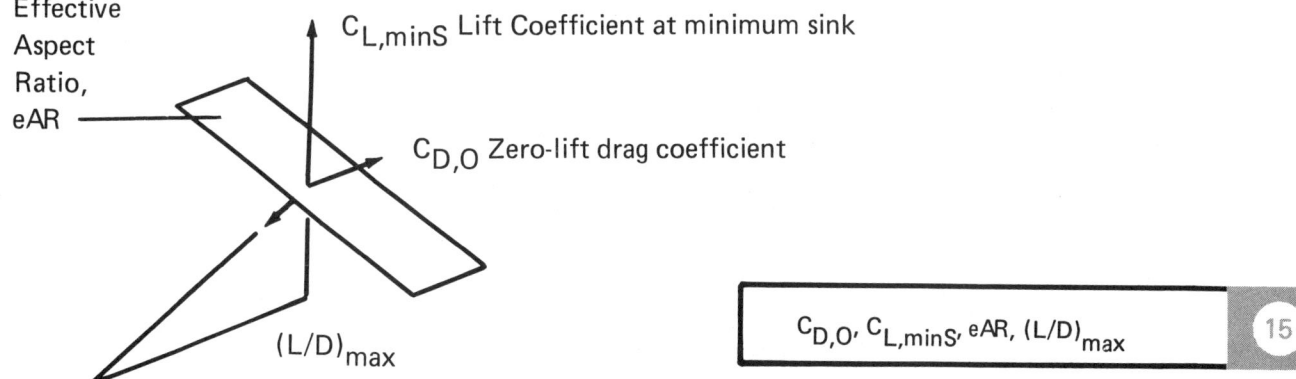

For reference: $C_{L,maxLD} = 0.577 \; C_{L,minS}$; $C_{D,minS} = 4 \; C_{D,O}$; $C_{D,maxLD} = 2 \; C_{D,O}$; $(L/D)_{minS} = 0.866 \; (L/D)_{max}$.

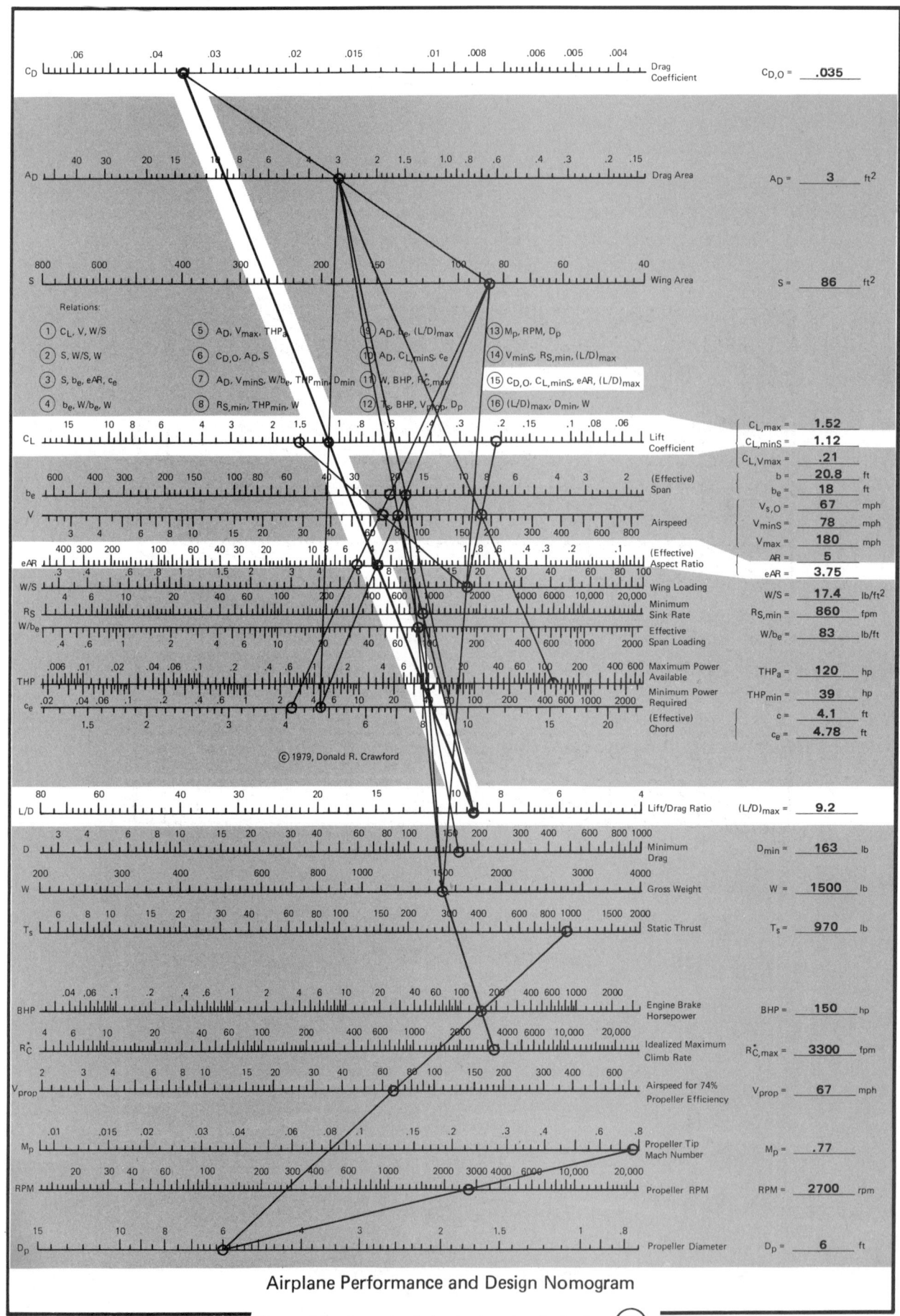

Figure 32. Sample Calculation for Relation (15).

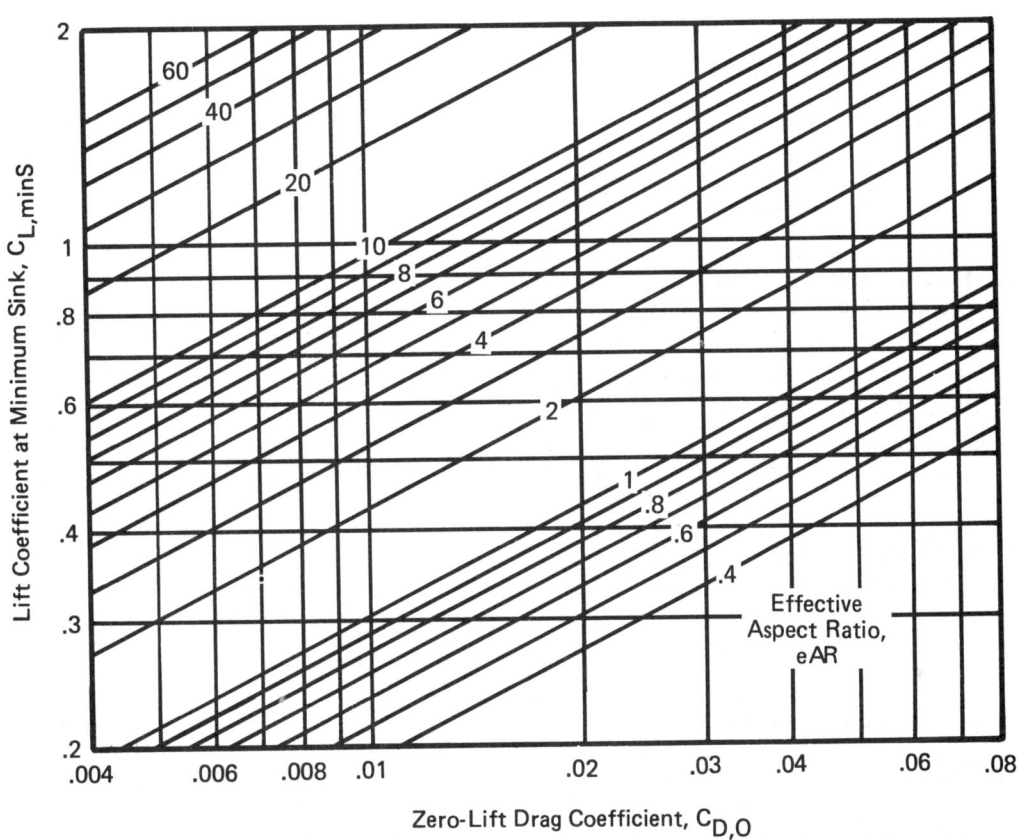

Figure 33. Relation ⑮ : Zero-Lift Drag Coefficient, Lift Coefficient at Minimum Sink, Effective Aspect Ratio. $C_{L,minS} = 3.07 \sqrt{eAR\, C_{D,O}}$

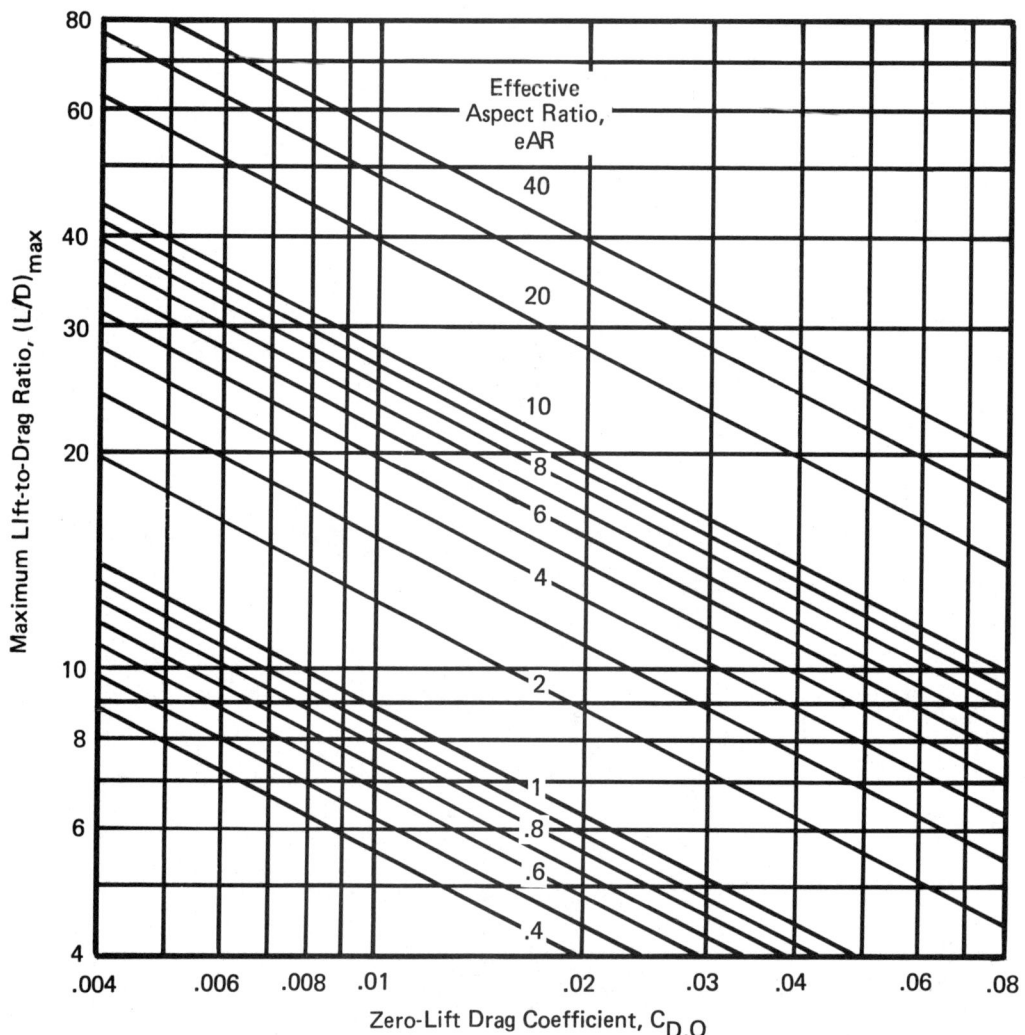

Figure 34. Relation ⑮ : Zero-Lift Drag Coefficient, Effective Aspect Ratio, Maximum Lift-to-Drag Ratio.
$$(L/D)_{max} = 0.8862 \sqrt{eAR/C_{D,O}}.$$

Relation ⑯ : Maximum Lift-to-Drag Ratio, Minimum Drag, Gross Weight

This is another cross check of the graphical calculation procedure. We have already calculated the maximum lift-to-drag ratio from ⑨ and the minimum drag force from ⑦. If these values are connected on the nomogram in Figure 35, we check with the assumed value for the weight on the W-scale. In the sample calculation for the T-18, $(L/D)_{max}$ = 9.2, D_{min} = 163 lb and W = 1500 lb. These are consistent results as can be seen from the straight line relation on the figure.

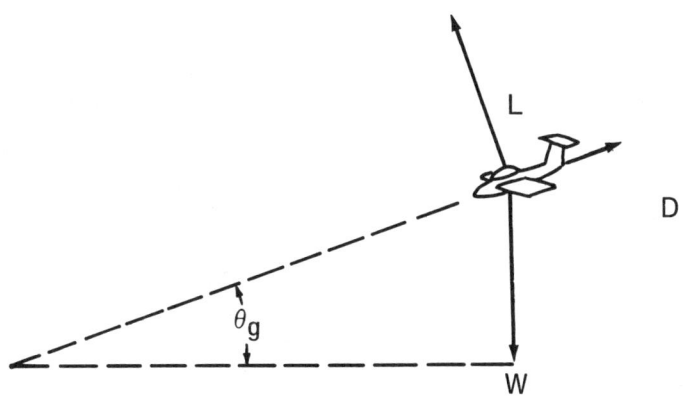

$$\theta_{g_{min}} = \frac{180}{\pi} \frac{1}{(L/D)_{max}}$$

Minimum glide angle (degrees) corresponds to maximum L/D.

$(L/D)_{max}$, D_{min}, W   16

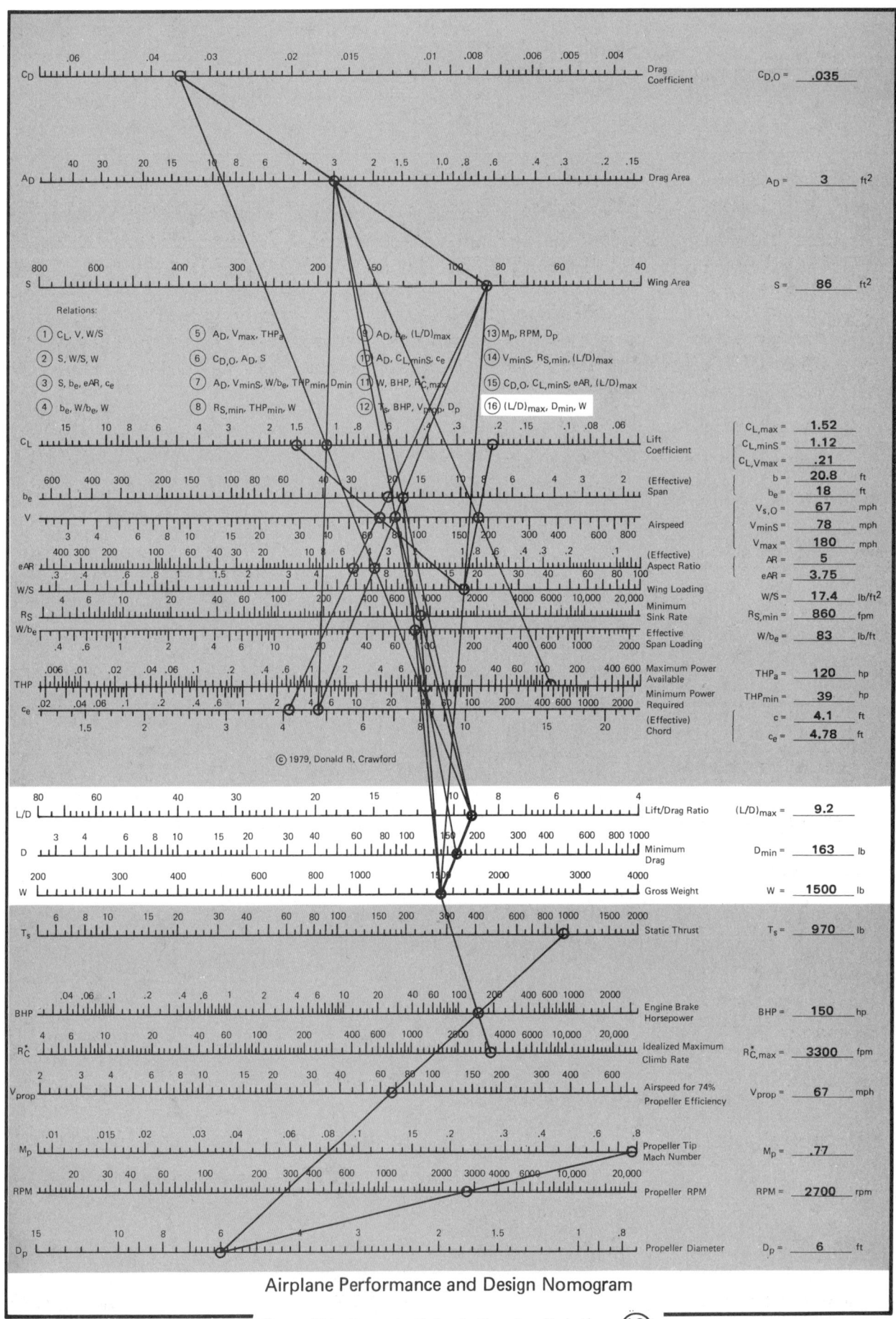

Airplane Performance and Design Nomogram

Figure 35. Sample Calculation for Relation ⑯.

SUMMARY OF THE NOMOGRAM CALCULATION FOR THE T-18

In the course of the sample calculation for the T-18, each new line was added in succession to the nomogram. The completed calculation on the Airplane Performance and Design Nomogram is shown in Figure 36, where we can see that the cross checks give us confidence in the graphical calculation. The results are tabulated at the right hand side of the nomogram, in the spaces provided, and are summarized below. We are now in a position to use the plastic template to find the rate of climb as a function of airspeed. The important results that we need from the nomogram to do this are: $R_{S,min}$, the minimum sink rate; V_{minS}, the airspeed for minimum sink; $R^*_{C,max}$, the maximum ideal climb rate (equal to the rate the engine can lift a weight equal to that of the airplane); and V_{prop}, the reference airspeed which would result in an ideal propeller efficiency of 74%.

Input quantities:

Stall speed	V_s = 67 mph	
Maximum Lift Coefficient	$C_{L,max}$ = 1.52	
Gross weight	W = 1500 lb	
Wing span	b = 20 ft 10 in	
Maximum level flight speed	V_{max} = 180 mph	
Engine brake horsepower	BHP = 150 hp	} max available
Assumed propeller efficiency	η = 0.8	} thrust horsepower
Propeller diameter	D = 6 ft	} THP = 120 hp
Propeller rotational speed	RPM = 2700 rpm	

Calculated quantities:

Wing loading	W/S = 17.4 lb/ft^2	Relation ①
Wing area	S = 86 ft^2	Relation ②
Aspect ratio	AR = 5	Relation ③
Wing chord	c = 4.1 ft	Relation ③
Airplane efficiency factor	e = 0.74	Appendix F
Effective aspect ratio	e AR = 3.7	Relation ③

Effective span	b_e = 18 ft	Relation ③
Effective chord	c_e = 4.75 ft	Relation ③
Effective span loading	W/b_e = 83 lb/ft	Relation ④
Drag area	A_D = 3 ft²	Relation ⑤; Appendix G
Zero-lift drag coefficient	$C_{D,O}$ = 0.035	Relation ⑥
Airspeed for minimum sink	V_{minS} = 78 mph	Relation ⑦
Minimum required power for level flight	THP_{min} = 39 hp	Relation ⑦
Minimum drag	D_{min} = 163 lb	Relation ⑦
Minimum sink rate	R_{Smin} = 860 fpm	Relation ⑧
Maximum lift-to-drag ratio	$(L/D)_{max}$ = 9.2	Relation ⑨
Lift coefficient at minimum sink	$C_{L,minS}$ = 1.12	Relation ⑩
Max ideal rate of climb	$R^*_{C,max}$ = 3300 fpm	Relation ⑩
Speed for 74% propeller efficiency	V_{prop} = 67 mph	Relation ⑪
Ideal static thrust	T_s = 970 lb	Relation ⑫
Propeller tip Mach number	M_p = 0.7	Relation ⑬

A tablet of Airplane Performance and Design Nomograms is included with the present book. These can be used for making your own preliminary design calculations by following the step by step procedures outlined in the text. A few sheets of 2 cycle x 2 cycle logarithmic graph paper are also included so that the plastic template can be used to make the rate of climb calculations described in the next section.

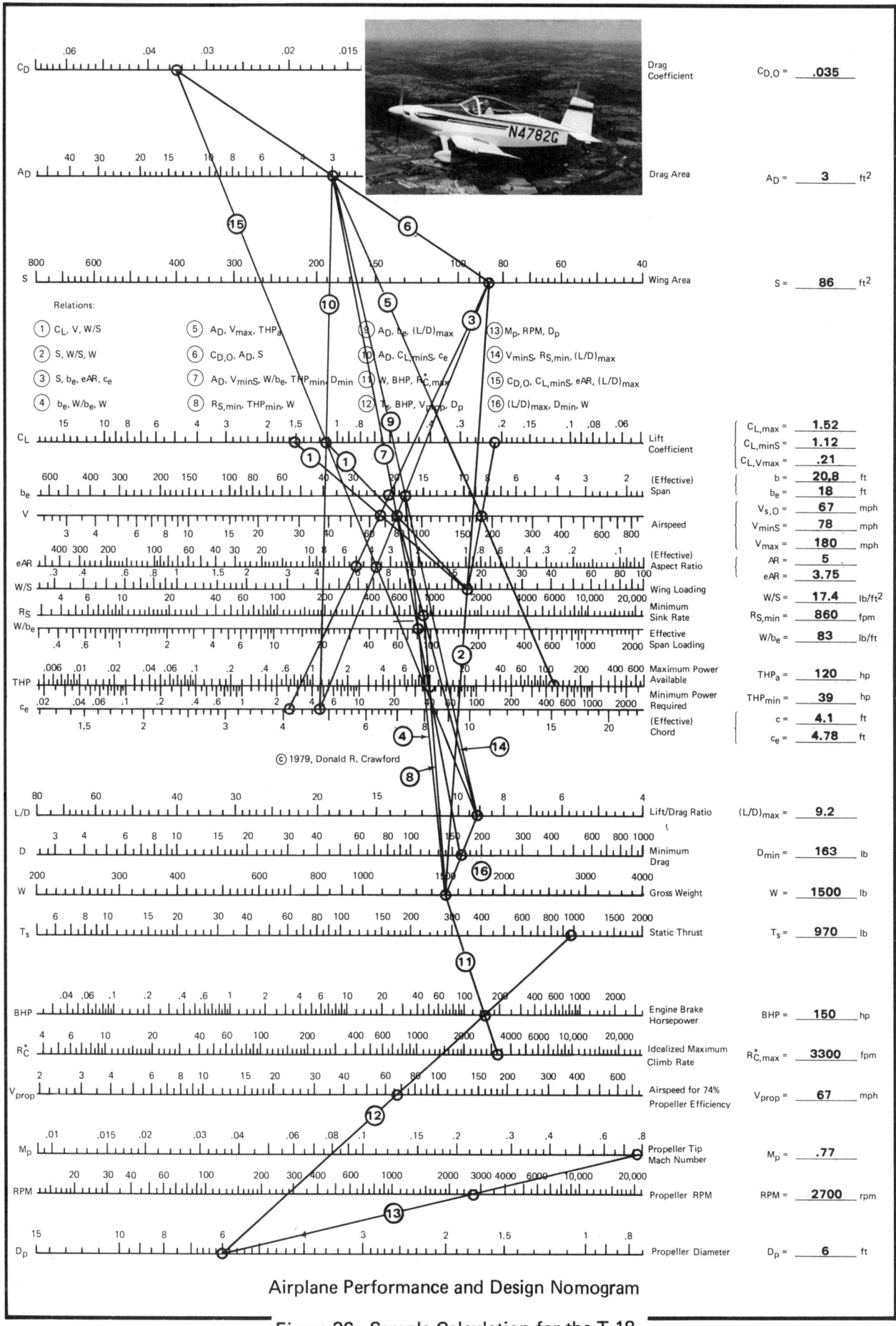

Figure 36. Sample Calculation for the T-18.

HOW TO USE THE TEMPLATE FOR CALCULATION OF RATE OF CLIMB

In Part 2 we show that the rate of climb, R_C, can be calculated from the sink rate, R_S, the maximum ideal climb rate, $R^*_{C,max}$, and the propeller efficiency, η. This is expressed mathematically by equation (38) or

$$R_C = R^*_{C,max} \eta - R_S$$

where $R^*_{C,max}$ is calculated with the nomogram from relation ⑪. The propeller efficiency and the sink rate are functions of the forward airspeed and are related to the nondimensional curves given by Figures 37 and 38. The nondimensional curves have been made into a plastic template that can be used to draw the correct curves for the sink rate and ideal climb rate versus airspeed, once we know where to position the template. The template is designed for use with 2 cycle x 2 cycle logarithmic graph paper 7.5 inches on a side (for instance, K&E NO. 46 7203 or Dietzgen No. 340R-L22). **WARNING:** The use of any other paper size will result in the wrong answer for sink rate and climb rate!

The construction of the curves follows a straightforward procedure:

1. Take a piece of 2 cycle x 2 cycle logarithmic graph paper and label the axes: rate of climb on the vertical axis and airspeed on the horizontal axis.

2. From the nomogram, mark the reference point for the sink rate curve. That is, place a cross at the point (V_{minS}, $R_{S,min}$). In the sample calculation V_{minS} = 78 mph and $R_{S,min}$ = 860 fpm).

3. Take the template for the sink rate curve and overlay on the graph paper, taking care to line up the reference point on the paper with the reference point on the template.

4. Draw the curve for the sink rate as a function of the airspeed using the template as if it were a French curve (Figure 39 for the T-18 calculation.)

5. From the nomogram mark the reference point for the propeller efficiency curve. The point is determined by the point (V_{prop}, $R^*_{C,max}$). In the sample calculation for the T-18, V_{prop} = 67 mph and $R^*_{C,max}$ = 3300 fpm.

6. Take the template for the efficiency (rate of climb) and overlay it onto the graph paper, aligning the reference point on the template with the correct reference point on the graph paper.

7. Draw the curve for the ideal rate of climb as a function of the airspeed in a manner similar to step 4. The curve drawn with the template corresponds to 85% of the ideal efficiency for the engine propeller combination. (See Figure 39 for the T-18.)

8. In order to find the rate of climb at a particular airspeed, subtract the sink rate from the ideal climb rate at that speed. This can be plotted on a separate plot with linear scales as in Figure 40.

9. The maximum level speed is determined directly from the logarithmic graph where the two lines cross (Figure 39). This will also correspond to the point on the rate of climb curve (Figure 40) where the rate of climb is zero. For the sample calculation the maximum level flight speed is 175 mph.

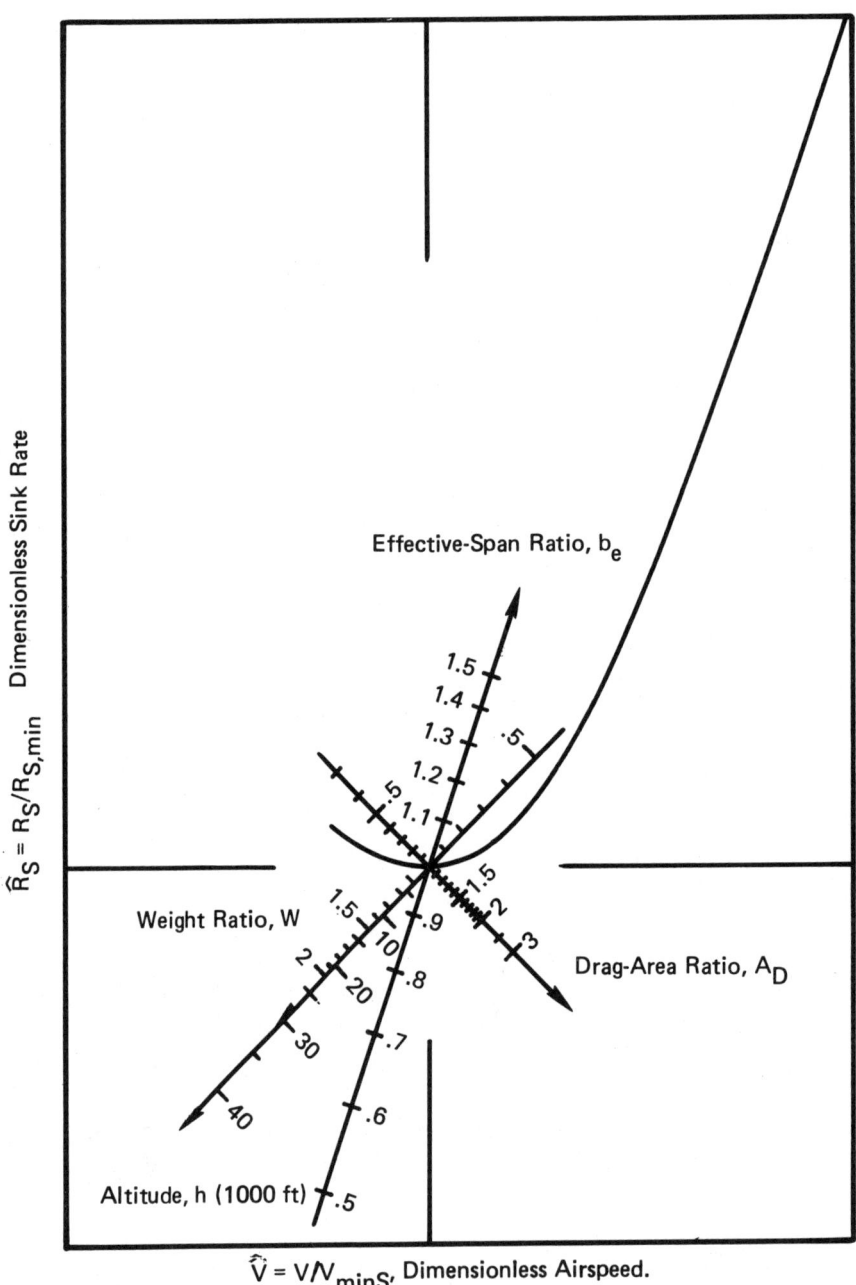

Figure 37. Dimensionless Sink Rate versus Airspeed with Parametric Effects of Drag Area, Effective Span, Weight and Altitude. $\hat{R}_S = (\hat{V}^4 + 3.)/4\hat{V}$.

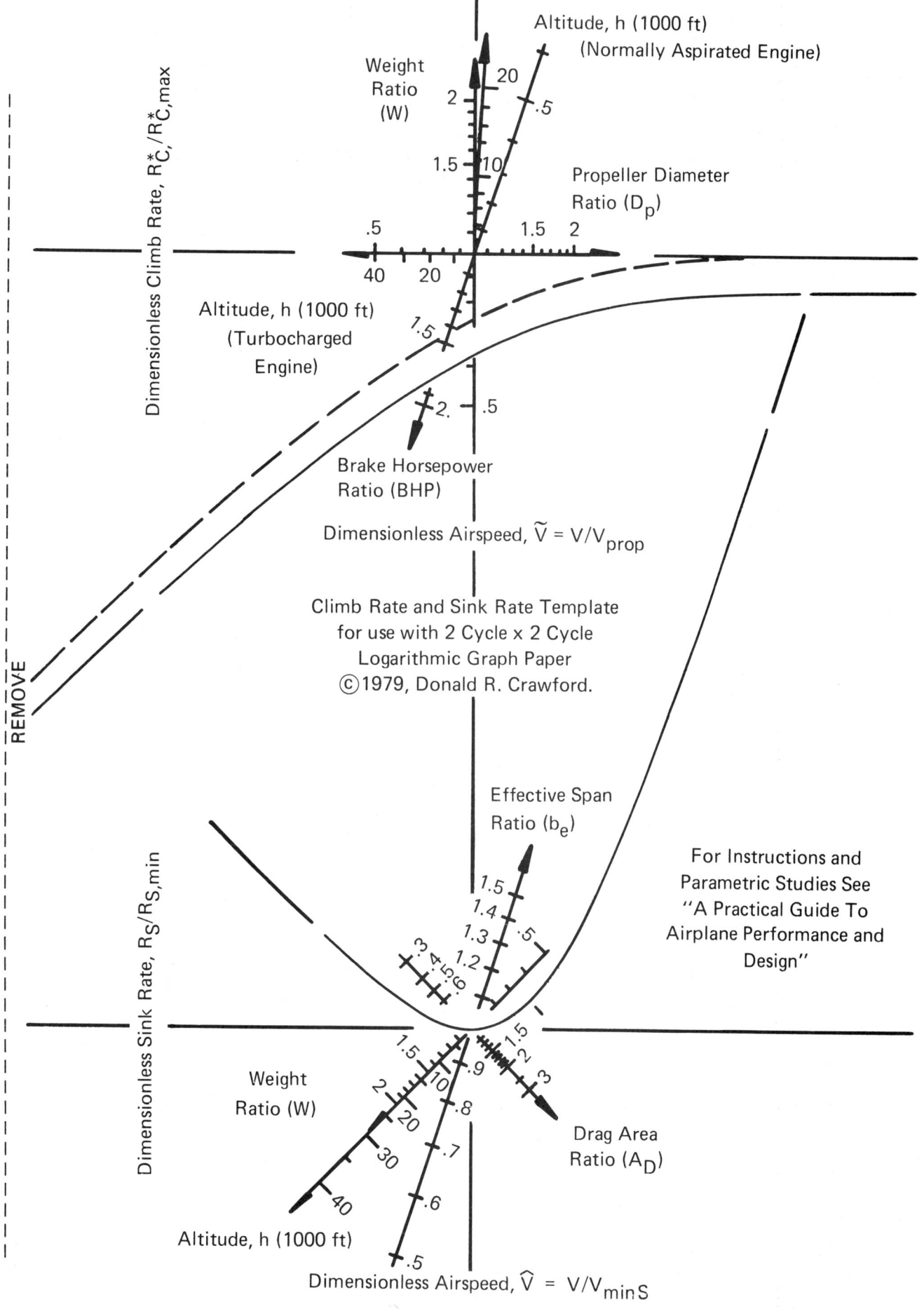

REMOVE

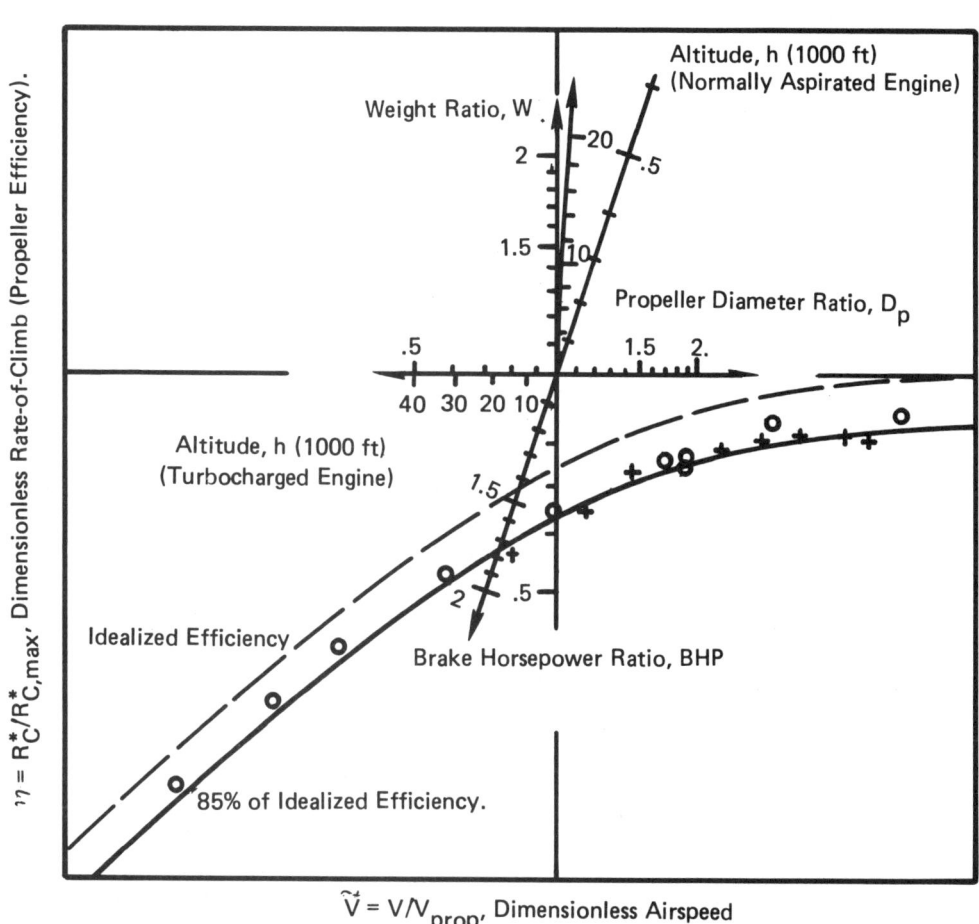

Figure 38. Dimensionless Idealized Rate-of-Climb versus Airspeed with Parametric Effects of Brake Horsepower, Propeller Diameter, Weight and Altitude.

In Figure 39 we have plotted the sink rate and the ideal rate of climb as functions of the relative airspeed. The ideal climb rate decreases with decreasing airspeed because the propeller becomes less efficient at low speeds. The sink rate has a minimum of 860 fpm at an airspeed of 78 mph and then increases rapidly at higher airspeeds. For high speeds the sink rate is proportional to the cube of the airspeed. When the sink rate and the ideal climb rate are equal, we find the maximum level flight speed — V_{max} = 175 mph. At lower speeds we have an excess of power available and we are able to climb.

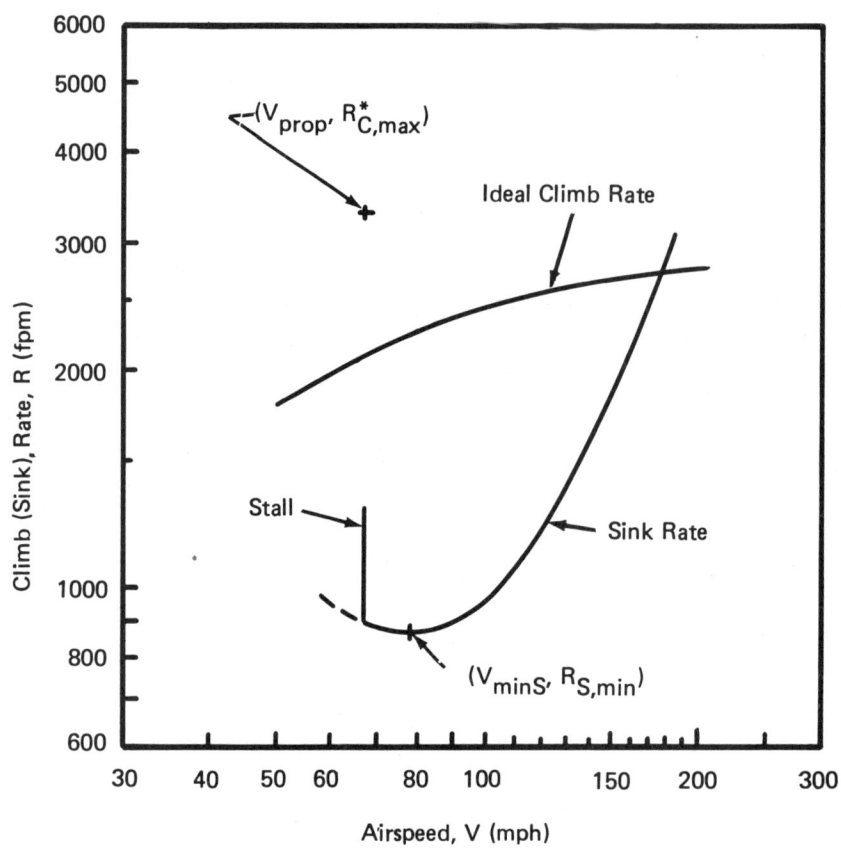

Figure 39. Sample Calculation of Ideal Climb Rate and Sink Rate for the T-18.

The climb rate can be found directly from Figure 39 by subtracting the sink rate from the ideal climb rate. If this is plotted in Figure 40 on a more familiar linear scale, we see that the maximum rate of climb, 1500 fpm, occurs at a speed of about 98 mph. The airspeed for best rate of climb is called V_y.

In the stall analysis discussion (see relation ①) we assumed that the stall speed, $V_{s,0}$, was 67 mph. This speed corresponds to the maximum lift coefficient that the airfoil can produce. The theory developed in Part 2 assumes that the lift coefficient varies linearly with the angle of attack. But, from the airfoil data in Appendix H, we see that the linear range of the lift coefficient extends up to about C_L = 1.4. From relation ① on the nomogram we find that this corresponds to about 70 mph, where there will be a rapid transition to the stall condition and a sudden decrease in the climb rate curve. Since this is such a sharp change, we can simply modify the shape of the curve in this narrow region between 67 and 70 mph by eye or by using a French curve. This will give the character of the results in the vicinity of the stall speed.

For short field takeoffs, we try to fly at an airspeed that gives the best climb angle, $\theta_{C,max}$. This speed is called V_x and is found from the curve of climb rate versus airspeed in Figure 40 by drawing the line through the origin tangent to the rate of climb curve. The airspeed for best angle of climb is about 75 mph. Although the climb rate, 1310 fpm, at this speed is less than the maximum climb rate, the climb angle is greater. The climb angle can be found from the equation

$$\theta_C = \frac{180}{\pi\, 88} \frac{R_C}{V}$$

where R_C is the rate of climb in feet per minute and V is the airspeed in miles per hour. This is an approximate equation valid for small angles (less than about 20 degrees). The maximum climb angle found from the tangent in Figure 40 is 11.4 degrees.

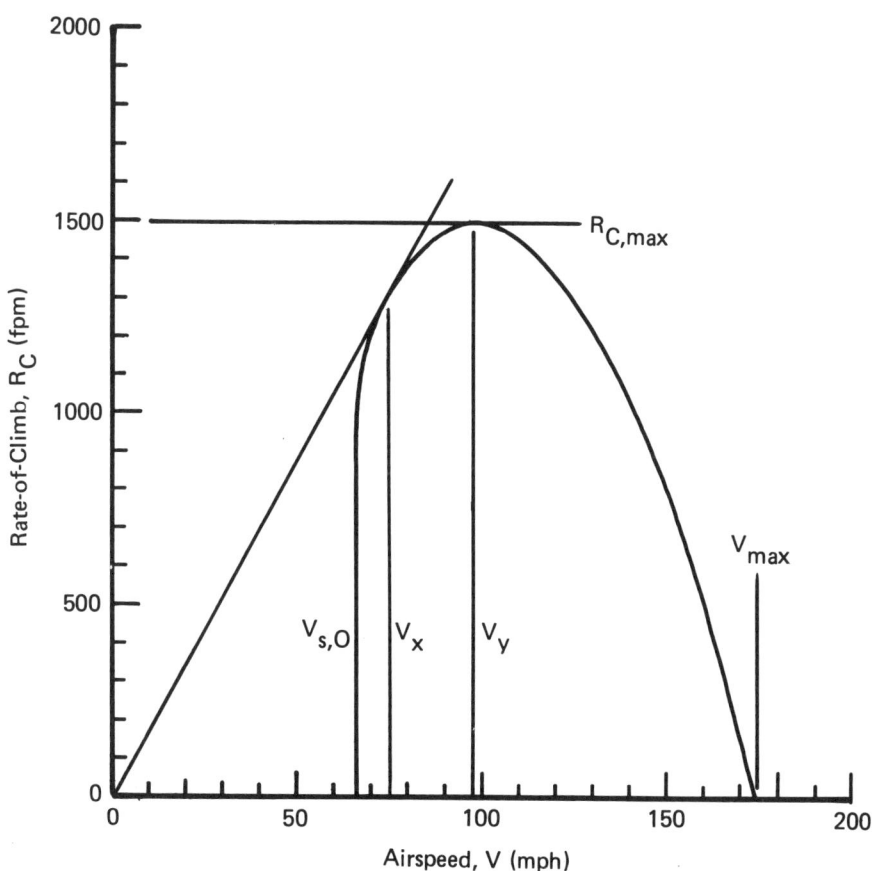

Figure 40. Rate of Climb versus Airspeed for Sample Calculation of Thorp T-18.

PARAMETRIC STUDY OF THE BASELINE DESIGN

Now that we have calculated the performance for the baseline design for the conditions of maximum engine horsepower at sea level, we would like to learn about the effects of each of the parameters on the design. For instance, if we change the power setting, the level flight speed will change. This effect of the power setting on the cruise speed is important when we want to find the cruise speed that will give the best economy. Also, since engines are typically run at 75% power, we can find the corresponding cruise speed.

Similarly, with an increase in altitude, the engine develops less power (unless it is turbocharged), and the sink rate curve shifts upward and to the right. Our climb rate, which is the difference between the two curves, decreases with altitude until we reach the absolute ceiling. We will show how to calculate this as well as the service ceiling (defined as the altitude at which the climb rate decreases to 100 fpm).

Then, we can look at the effects of leaving the passenger behind or taking too much baggage; streamlining to reduce drag; ground effect, etc. These effects can be quickly evaluated with the help of the plastic template — an overlay of Figure 37 and 38. The template has scales that show where the location of the reference point will move when we make changes in a parameter. When the template is properly positioned, we can draw the sink rate and ideal climb rate curves. For instance, if we increase the weight, we expect that the sink rate will increase and the ideal climb rate will decrease. For the sink rate template, the arrow for changes in weight points down and to the left at a 45 degree angle. This direction results from the fact that both the minimum sink rate and the speed for minimum sink increase with increasing weight. Altitude has a similar effect on the location of the sink rate curve. On the other hand, changes in the drag area cause the new location of the sink rate curve to be positioned in a different direction. That is, an increase in the drag area increases the sink rate, but **decreases** the speed for minimum sink. This results in an arrow for drag area changes that points in the 45 degree direction down and to the right on the plastic template. When the template is placed on the 2 cycle x 2 cycle logarithmic graph paper, and the reference point is aligned for the new drag condition (say the drag is increased to a value 1.2 times the original value) then the curve will be slightly displaced from the baseline design in a direction up and to the left. This will result in a decrease in the maximum speed if the power available remains constant. This effect and other parametric variations will be discussed in the following sections.

POWER SETTING, CRUISE SPEED, RANGE

In the baseline calculation we assumed values for the maximum brake horsepower conditions at sea level to find the rate of climb as a function of airspeed and to find the maximum airspeed. We would now like to perform a parametric analysis to find the dependence of the sea level cruise airspeed on the power setting. The template that was developed for use in the ideal rate of climb can be used directly to determine the effects of power-setting. This is done by moving the template so that the reference point is aligned with each new power setting. The sample calculation plotted in Figure 41 was made by placing the reference point at brake horsepower ratios of 1.0 (the original full-throttle case), 0.9, 0.8, 0.7, 0.6, 0.5 and 0.4. In this succession of plots the curve for the ideal available climb rate moves down and to the left at a ratio of 3:1. This ratio follows from the expression for the V_{prop} (see relation ⑫).

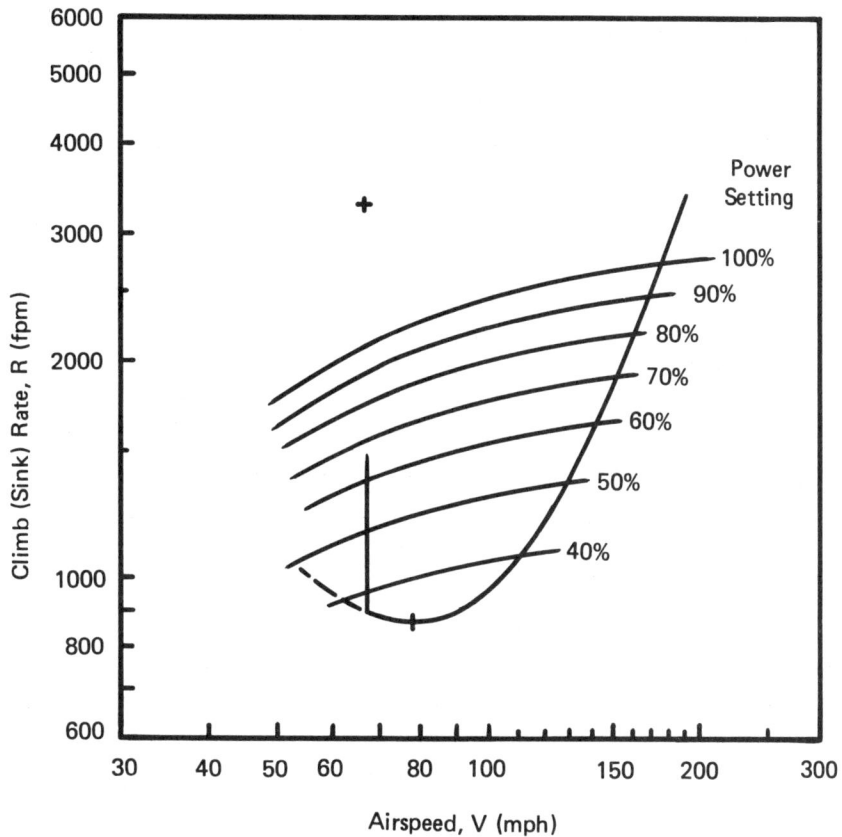

Figure 41. Effect of Power Setting on Performance.

When the BHP is plotted versus the cruise speed (found from the intersection of the curves on Figure 41), we obtain Figure 42. The theoretical value of the thrust horsepower required for level flight is also plotted. We can get an idea of the propeller efficiency from this figure since the efficiency, η, is the ratio between the thrust horsepower and the brake horsepower. Also, we can find the airspeed that gives the best range from the tangent to the BHP-V_c curve. This is the speed that would theoretically yield the best ratio of speed-to-horsepower, which translates into the speed for best range since less gas is burned per mile traveled. However, at the speed of 111 mph the engine is producing less than 40 percent of its maximum rated power. The engine manufacturers recommend that the engines should be run at 50 percent power or better, so the minimum economy cruise setting (50%) would give a cruise speed of 128 mph. At 75% power, the recommended power setting for continuous cruise conditions, the cruise speed would be 156 mph.

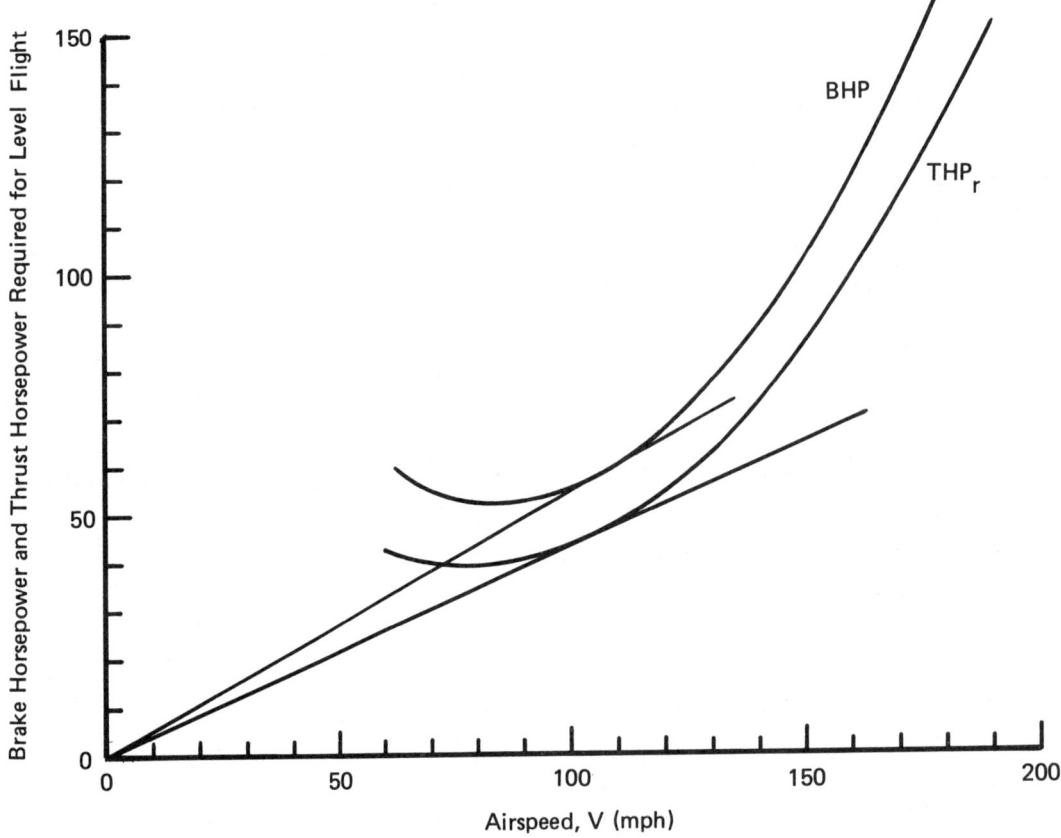

Figure 42. Brake Horsepower and Thrust Horsepower Required for Level Flight as Functions of the Airspeed.

If the T-18 is assumed to have a gas tank that holds 29 gallons of gasoline; the specific fuel consumption of the engine, SFC, is assumed to be .5 lb-fuel per horsepower per hour; gasoline weights 6 lb per gallon; and the power setting for economy cruise is 50% at a cruise speed of 128 mph, then the range is

$$\text{Range} = \frac{V_c \; 6 \; \text{Gal}}{\text{SFC} \; \text{BHP}} = \frac{128 \; (6) \; (29)}{(.5) \; (.5 \times 150)} = 594 \text{ miles}$$

If we calculate the range at 75% power (156 mph cruise speed), we get 482 miles. These calculations are comparable with the published range of the 180-hp T-18, which is 500 miles.

ALTITUDE EFFECTS; ABSOLUTE AND SERVICE CEILINGS

An increase in altitude affects the performance of the airplane in three ways. First, it reduces the available brake horsepower of the engine. This happens because the fuel-air mixture increases as a result of the decreased density of the outside air. If the over-rich mixture is leaned by reducing the gas flow rate, less fuel is available to the engine, and the power decreases. The only way to get around this is to turbocharge the engine so that more air is stuffed into the cylinders and the fuel-air mixture remains constant. The effectiveness of the turbocharger is limited, however, and above a certain critical altitude, the turbocharged engine also begins to lose power.

The second effect is the reduction in propeller efficiency. Even if the brake horsepower at the shaft of the engine remains constant, the propeller becomes less and less efficient as the altitude increases. This is modeled by the density ratio factor in the reference propeller speed, V_{prop}, as discussed in Part 2.

The third effect is the change in the sink rate curve. An increase in altitude has exactly the same effect on the sink rate curve as an increase in weight. That is, the reference point is shifted up and to the right with an increase in altitude. For low speeds, this results in an increase in the sink rate, but at high speeds this gives a relative decrease in the sink rate compared to the sea level value at this speed. All three of these effects are taken into account in the analysis of this section.

For normally aspirated engines, the power decreases with the decrease in density ratio. According to the discussion by Von Mises (1945) the brake horsepower at a density altitude corresponding to the density ratio, σ, is

$$BHP = \phi(h) \, BHP_{SL}$$

where the power-altitude factor is given by

$$\phi(h) = \frac{\sigma - c}{\sigma_1 - c} \, .$$

where the coefficient σ_1 is equal to 1.0 for normally aspirated engines and the value of c is 0.15. More recent data suggests that the value of c should be taken to be 0.12. For supercharged engines below the critical altitude, $\phi(h) = 1.0$, and for altitudes above the critical altitude, σ_1 corresponds to the density ratio at the critical altitude. The brake power-altitude factor, $\phi(h)$, is plotted in Figure 43, along with the variation of the density ratio, σ, versus the density altitude, h. The variation of the density ratio in the standard atmosphere is discussed in more detail in Appendix D.

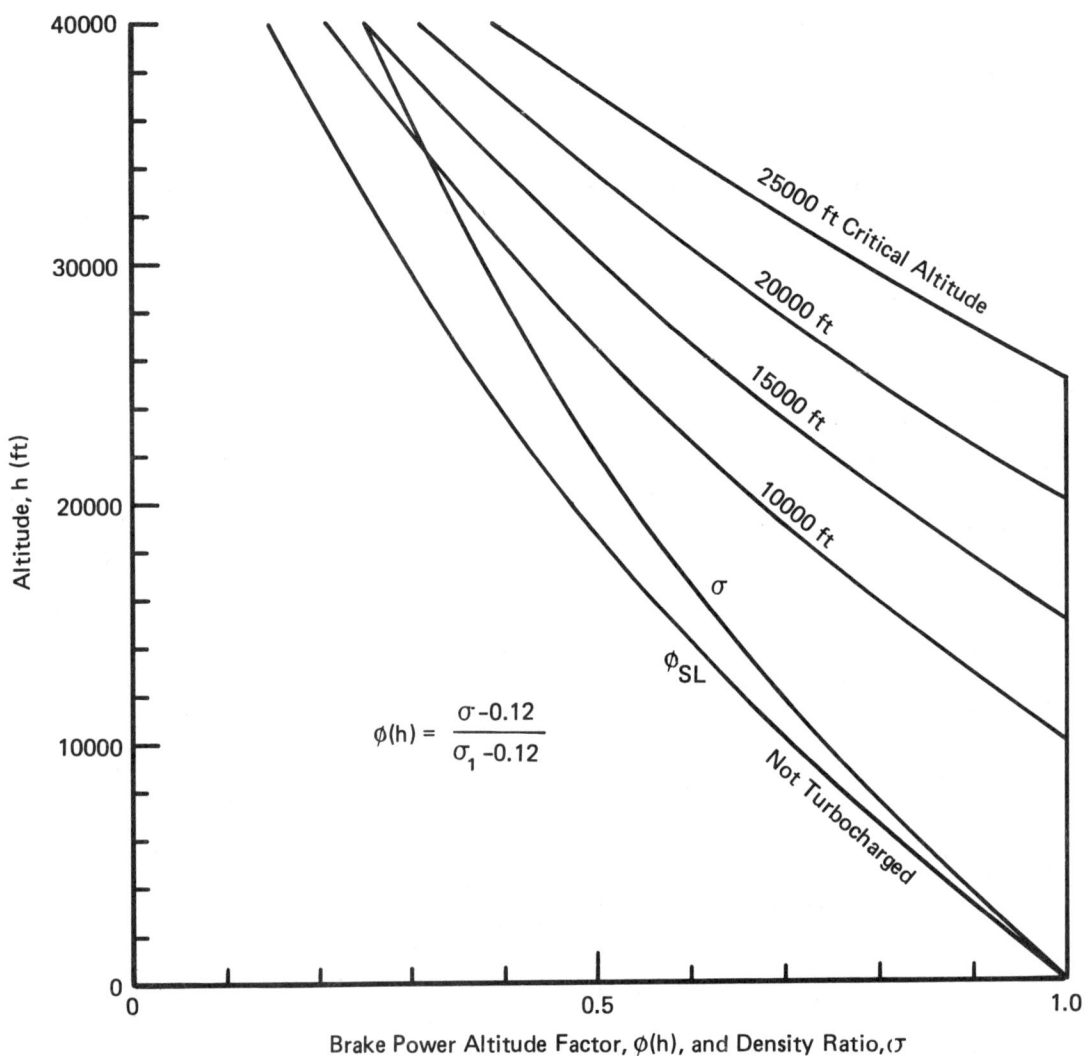

Figure 43. Brake Power-Altitude Factor, $\phi(h)$, and Density Ratio, σ, versus Altitude for Various Critical Altitudes for the Turbocharger.

The location of the reference point for the ideal climb rate curve, moves down and slightly to the left for normally aspirated engines, and to the right for turbocharged engines (below the critical altitude) as the altitude increases. This is demonstrated in Figure 44 for a normally aspirated engine and in Figure 45 for a turbocharged engine with a 10000 foot critical altitude.

The change in the sink rate curve is the same for the two cases. That is, the reference point for the minimum sink rate moves up and to the right with increasing altitude. When the rate of climb is determined for each of these cases by subtracting the sink rate from the ideal rate of climb, we obtain the rate of climb curves shown in Figures 46 and 47.

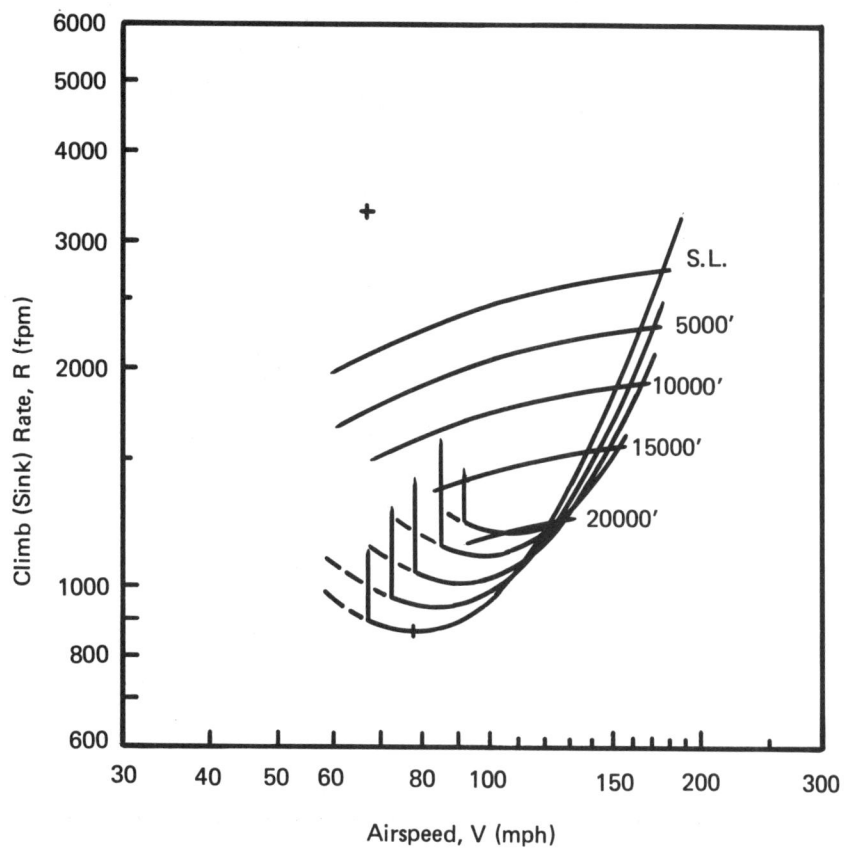

Figure 44. Effect of Attitude on Performance with Normally Aspirated Engine.

Consider first the normally aspirated engine in Figures 44 and 46. The ideal rate of climb decreases with altitude and the sink rate increases. The difference between the ideal rate of climb curve and the sink rate curve decreases so that the rate of climb always decreases with altitude. At the absolute ceiling the maximum rate of climb decreases to zero. At each altitude there is a variation of the climb rate with airspeed. The speed for maximum rate of climb, V_y, increases with altitude until the climb rate decreases to zero at the absolute ceiling. The speed for the best angle of climb also increases with altitude. At the absolute ceiling these two speeds approach each other. The maximum level flight speed however, decreases with altitude. The fastest speed for a normally aspirated engine is at sea level. Notice also that the stall speed increases with altitude because of the density ratio in the generalized form of relation ①(see Equation 7 in Part 2).

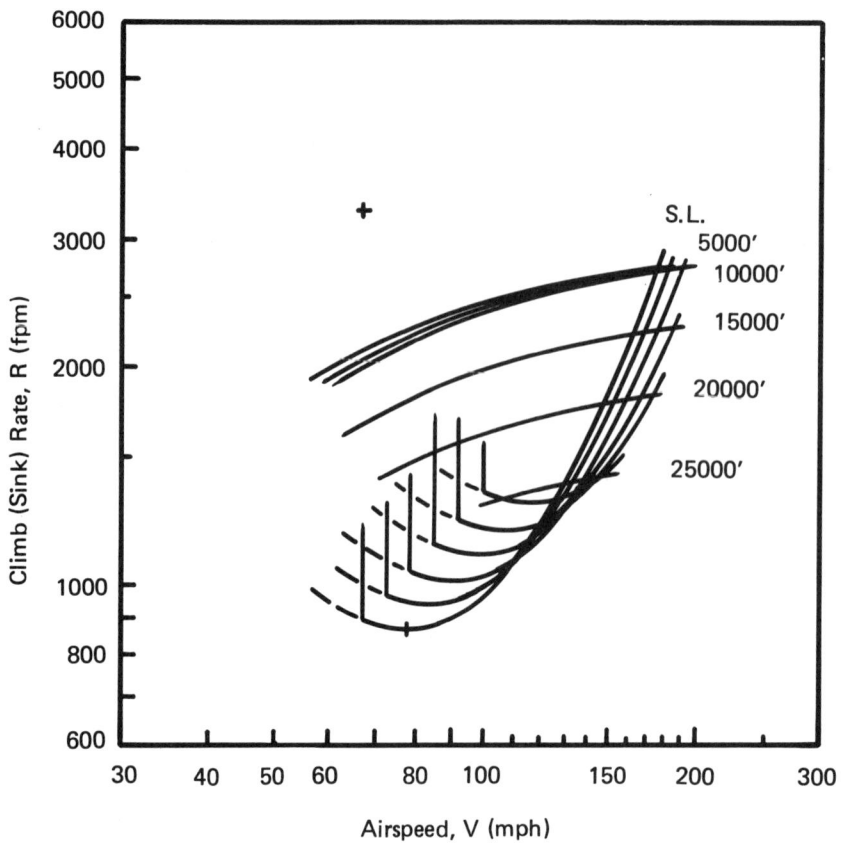

Figure 45. Effect of Attitude on Performance for a Turbocharged Engine with a 10,000 foot Critical Attitude.

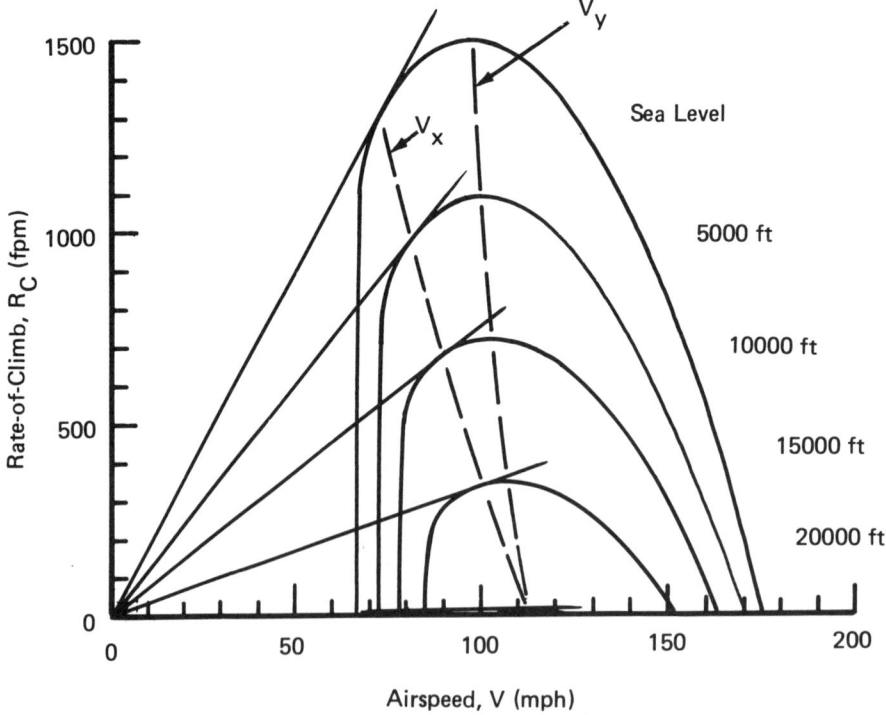

Figure 46. Effect of Altitude on Rate-of-Climb Performance for a Normally Aspirated Engine.

For the supercharged engine, we have similar curves in Figures 45 and 47. We have assumed that the supercharger has a critical altitude of 10,000 ft, so that below this altitude, we have virtually the sea level value for idealized rate-of-climb except for the propeller efficiencies. Above the critical altitude, the engine starts to lose power as in the case of the normally aspirated engine. Thus, the ideal rate-of-climb curve has two distinct characters above and below this altitude. This has a marked effect on the rate of climb curve in Figure 47. Until the supercharger starts to fade, the rate of climb is almost unaffected by altitude. In fact, the climb rates at the higher speeds are better than the sea level climb rates. Then, above the critical altitude, the rate of climb drops off in a similar manner as before. The curves for V_x and V_y have a kink in them at the 10,000 foot level — the turbocharger critical altitude. At the absolute ceiling, V_x and V_y are equal to each other, as before. The trend for the maximum level flight speed is different from that for the normal engine. The maximum level speed increases with altitude until the supercharger starts to fade, at which time the maximum level speed starts to decrease.

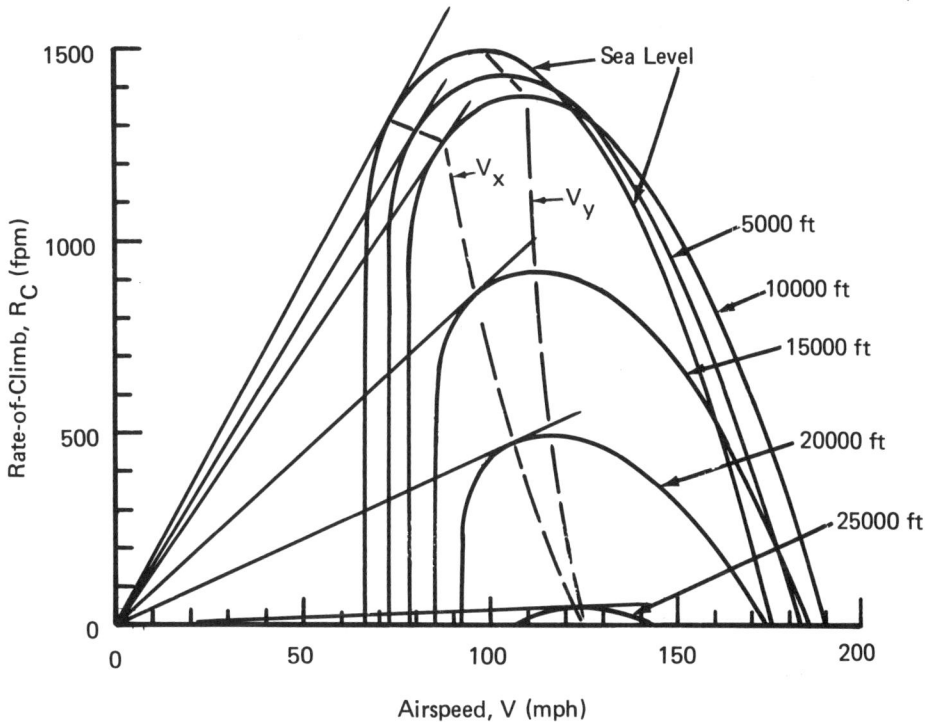

Figure 47. Effect of Altitude on Rate-of-Climb Performance for a Turbocharge Engine with a 10000 foot Critical Altitude.

If we take the maximum rate of climb for the normally aspirated engine and the turbocharged engine and plot them against the altitude, we obtain the curves in Figure 48. The scales are reversed so that altitude is the vertical axis. We see that the turbocharged engine always has a climb rate that is greater than the normal engine. However, they both start out with the same climb rate, unless the turbo is overboosted. At 10,000 feet the turbo starts to lose its effectiveness, and the rate of climb decreases, but the turbo still has more than twice the climb rate as that of the normal engine above 12,000 feet. This would be important when flying out of high altitude airports or when trying to avoid weather that may extend up to

20,000 feet. We have to remember, however, that above 14,500 feet we will be required to have oxygen according to the Federal Air Regulations (FARs). With this plot, we can extrapolate to find the absolute ceiling: 20,200 for the normally aspirated engine and 25,600 feet for the turbo. The service ceiling is defined as that altitude at which the rate of climb decreases to 100 feet per minute. For the normal engine it is 18,750 feet and for the turbo it is 24,500 feet. The published service ceiling for the 180-hp T-18 is 20,000 feet.

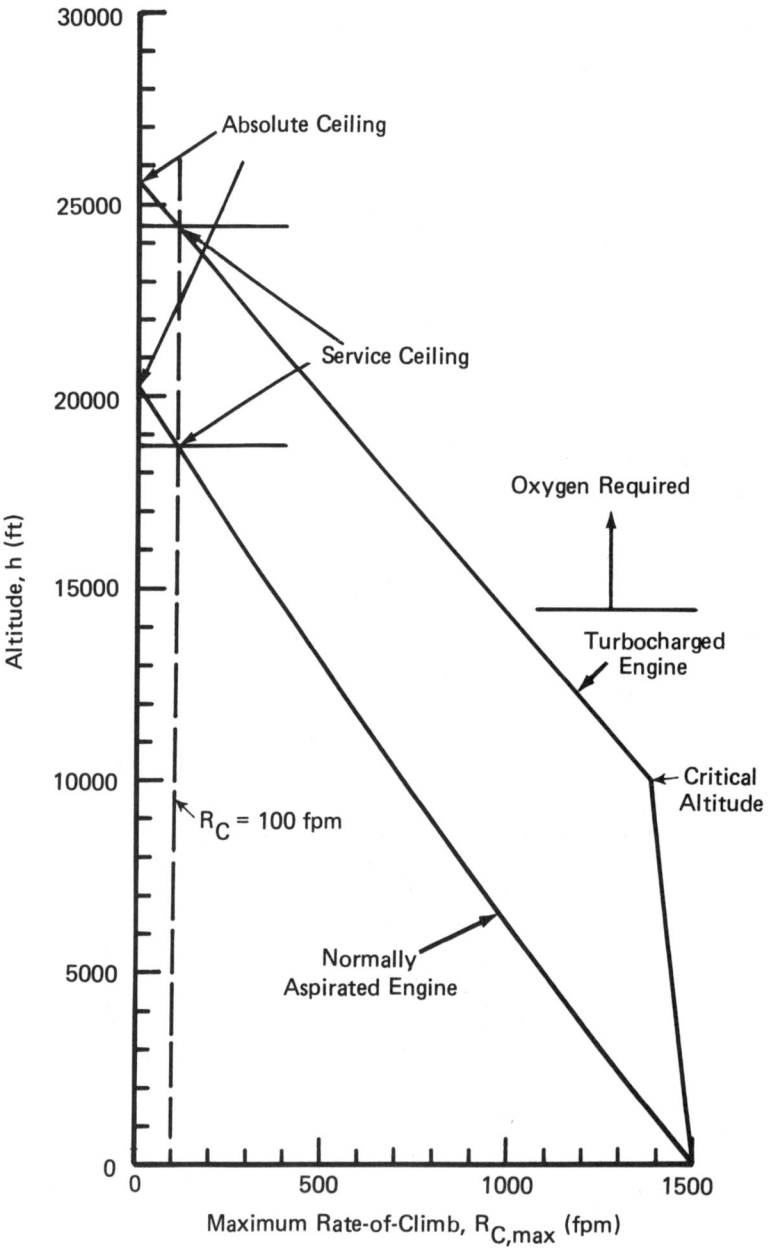

Figure 48. Dependence of Maximum Rate-of-Climb on Altitude. Definition of Service Ceiling and Absolute Ceiling.

If we plot the stall speed, V_s, the speed for maximum angle-of-climb, V_x, the speed for maximum rate-of-climb, V_y, and the maximum level flight speed, V_{max}, as function of the altitude, we obtain the curves in Figure 49. Notice the trends for the maximum speed for the two different types of engines. This is the reason we try to get to altitude as quickly as possible with the turbo — where it is more efficient. We see also why the racers of the normally aspirated engines try to race at the lowest possible altitudes where the higher power can be developed — unless there are more favorable winds at higher altitudes. The flight speeds are bounded at the high end by V_{max} and at the low end by V_s — the stall speed. This is the case until the airplane reaches the vicinity of the service ceiling. At this point, the minimum speed is governed by induced drag considerations and not the stall speed. That is, the airplane can be flown at a speed above the stall, but still not be able to climb. The airplane tends to mush along and loses altitude unless we fly at a speed above the minimum level flight speed.

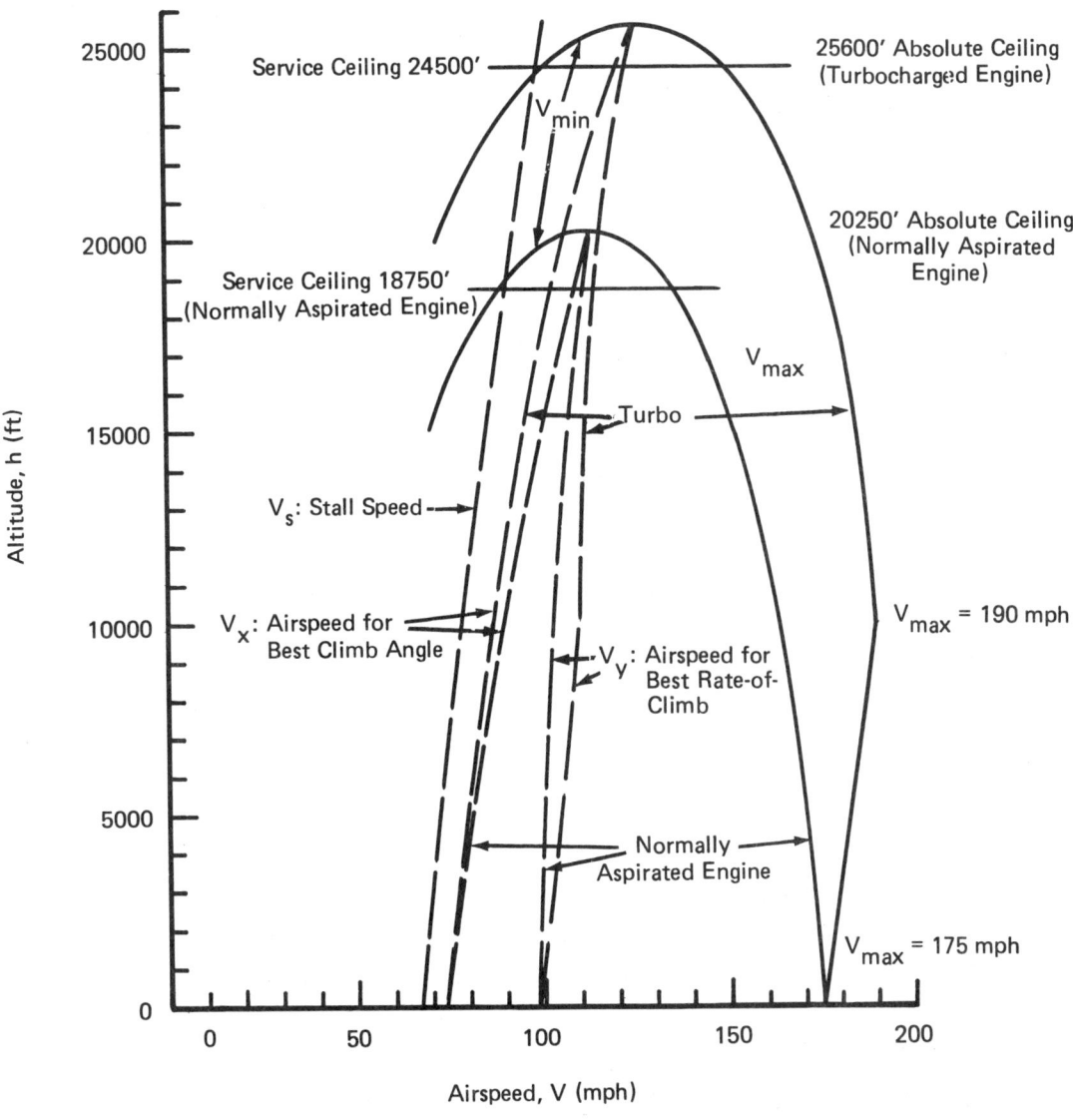

Figure 49. Effects of Altitude on V_s, V_x, V_y, and V_{max}.

WEIGHT EFFECTS

If the weight of the basic design is increased or decreased, we expect major effects upon the performance of the airplane. First, the ideal rate-of-climb is decreased because, for the same power input, the heavier weight is lifted at a slower rate. Also, the reference point for the sink rate curve moves upward and to the right for an increase in weight. This effectively increases the sink rate at lower speeds, but decreases the sink rate at the high speeds. That is the reason that the high performance cross-country sailplanes add ballast — to decrease the sink rate at high speeds between thermals (as long as the sink rate at low speeds will still allow soaring in the weak thermals).

Let us examine the effect of the weight by changing the weight of the basic design by plus or minus 10 percent — corresponding to 150 lbs too much baggage or one less passenger. The effect of the weight is easily taken into account by shifting the reference points, that is, by moving the template with respect to the originally determined reference points. For this parametric study, we use the weight arrow on the template. For the overweight condition, move the climb rate template down until the weight ratio of 1.1 is lined up with the reference point calculated from the nomogram (3300 fpm; 67 mph). Then, use the template like a french curve to draw the line for the ideal climb rate. If the weight decreases, we do the same procedure, except we move the template up until the reference point lines up with the 0.9 point on the weight-ratio line on the template. Draw the line using this specialized French curve to find that the ideal climb rate is higher than the basic case and the overgross case.

Next, we make similar modifications to the sink rate curve. The sink-rate template is first shifted up and to the right when the weight is increased, and then down and to the left when the weight is decreased. The scale for the weight ratio is used for this parametric study. The reference point has to be on this line, and the relative position of the template should be such that the horizontal and vertical axes are parallel to the axes on the 2 cycle x 2 cycle logarithmic graph paper.

The results in Figure 50 are plotted for the cases of an increase or decrease of 10 percent in the basic design weight. Note that the stall speed has changed because the wing loading has changed. This effect can be calculated from relation ① or by calculator, since the stall speed is proportional to the square root of the weight. Note also that the maximum level flight speed does not change very much, increasing slightly with decreasing weight. This follows from the approximation in relation ⑤, where the maximum speed depends on thrust horsepower and drag area, if we neglect induced drag. Since weight affects induced drag, we expect most effects at lower speeds. A discussion of the corresponding rate-of-climb (that is, the difference between the ideal rate of climb and the sink rate) will be deferred until we have considered all of the other parametric effects (Figure 54).

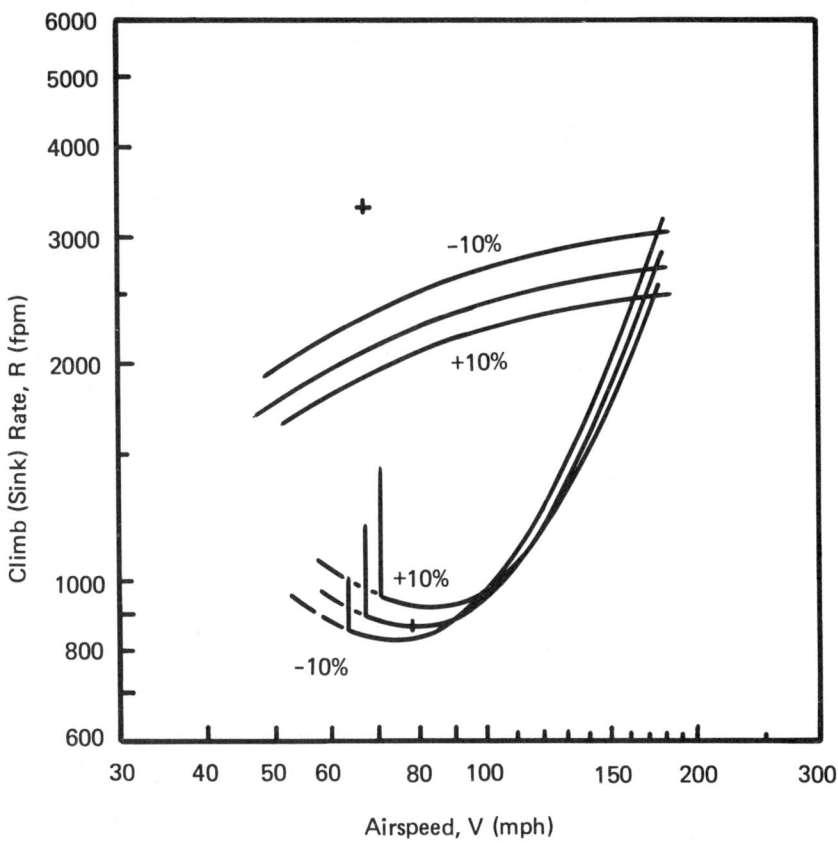

Figure 50. Effect of Weight on Performance.

DRAG EFFECTS: STREAMLINING AND FLAPS

Everyone knows that you are supposed to decrease drag if you want to increase performance. But how sensitive is maximum speed to the decrease in drag? And how much will the glide angle be affected by the drag of flaps? We can use our basic design and change the drag area to see the effects. First, we note that power is not affected, so the ideal rate-of-climb is the same for all three cases: baseline design, 10 percent decrease in drag area, and added flaps.

If we decrease the drag area by better streamlining on the wheels, washing and waxing the airplane skin, enclosing the communication and navigation antennas in the wing tips, making a good fitting cowling with the minimum cooling drag, etc., then we can move the sink rate template down and to the right. This will give us better climb performance, since there is more difference between the ideal climb-rate and the sink-rate curves, and a higher level flight speed, since the intersection of the two curves is farther to the right. The effect of a 10% decrease in drag area is shown in Figure 51. Note the increase in maximum speed — from 175 to 183 mph — for the 10% drag area decrease.

When flaps are deployed, several effects occur simultaneously. First, the lift coefficient increases dramatically. This causes a ballooning tendency until the airplane is slowed down or unless the angle of attack of the airplane is decreased. If the angle of attack were decreased until the lift coefficient is zero, we would find that the drag is larger than when the flaps are retracted. Therefore, we have increased the drag area. Finally, the maximum lift coefficient is much larger than the unflapped case, $C_{L,max}$ = 2.5 with flaps, compared to 1.52 without. But, since the flaps only extend over 60% of the wing span, the effective lift coefficient is about 2.1 (= .6 x 2.5 + .4 x 1.52). With this lift coefficient and the same wing loading as in the example, we can use relation ① on the nomogram to find that the stall speed with flaps is 57 mph compared to 67 mph with no flaps. The drag analysis of Appendix G shows that the drag area is increased by about 2 ft^2. And, because we started with 3 ft^2 of drag area, our new drag area is about 5 ft^2, or 1.67 times the original. The shift of the sink-rate template for the increased drag is up and to the left. That is, we have an increase in the sink rate, and a decrease in the speed for minimum sink — both desirable features for landing. Placing the template on the drag area-ratio point of 1.67 and aligning the axes so that they are parallel to the axes on the graph paper, we can draw the new sink-rate curve for the flaps-extended case on Figure 51. Note also, that the stall speed is decreased for this case since we have used the flaps to increase the lift coefficient of the wing.

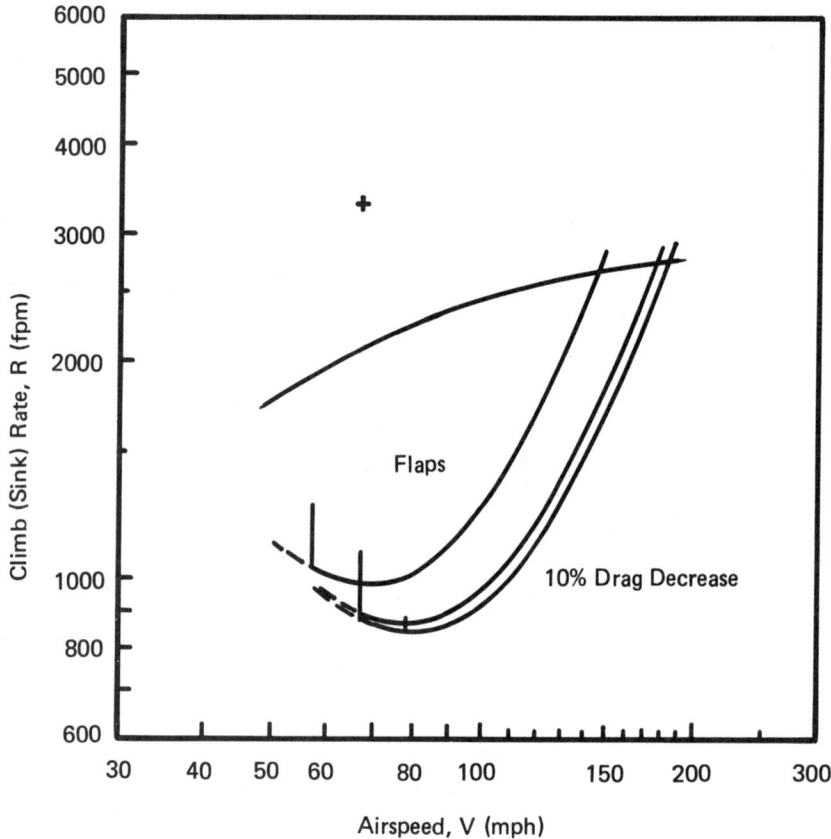

Figure 51. Effects of Streamlining and Flaps on Performance.

The effect of the addition of power on the climb rate with the flaps down (in the case where we have to "go-around") can be addressed by subtracting the sink-rate curve from the ideal climb-rate curve. A discussion of this, and the effect of streamlining on the climb rate will be deferred until rest of the parametric effects are discussed (Figure 54).

EFFECTIVE SPAN CHANGES; GROUND EFFECT

In this section, we would like to find the effect of changing the effective span of the basic airplane. First, consider the ground effect which essentially increases the airplane efficiency factor and so increases the effective span, $b\sqrt{e}$. If the airplane is flying within one wing span distance above the ground, we can find the effect on the airplane efficiency factor from Appendix F. Suppose that we are flying with no flaps at a height of 5 feet above the ground. Since the span is 20 feet 10 inches, the ratio of the height to the span is $h/b = 5/20.833 = 0.24$. From Figure F.3, we find that $k_{gd} = 1.31$, so that the effective span is increased by the factor $\sqrt{k_{gd}} = 1.15$. The effect on the baseline design can be found by shifting the sink rate template by the appropriate amount so that the reference point is aligned with the increased effective span ratio of 1.15. That is, the effective span has been increased by 15 percent. The sink-rate curve is then drawn on Figure 52, and compared to the baseline design. The biggest effect is at low speeds, where we have an increased rate-of-climb compared to what we would have if we were not in ground effect. The airplane has an increased efficiency because the ground effect has reduced the amount of induced drag. This is one reason for the success of the manpowered aircraft, since the power required for flight in ground effect at low altitudes is so much lower than at higher altitudes.

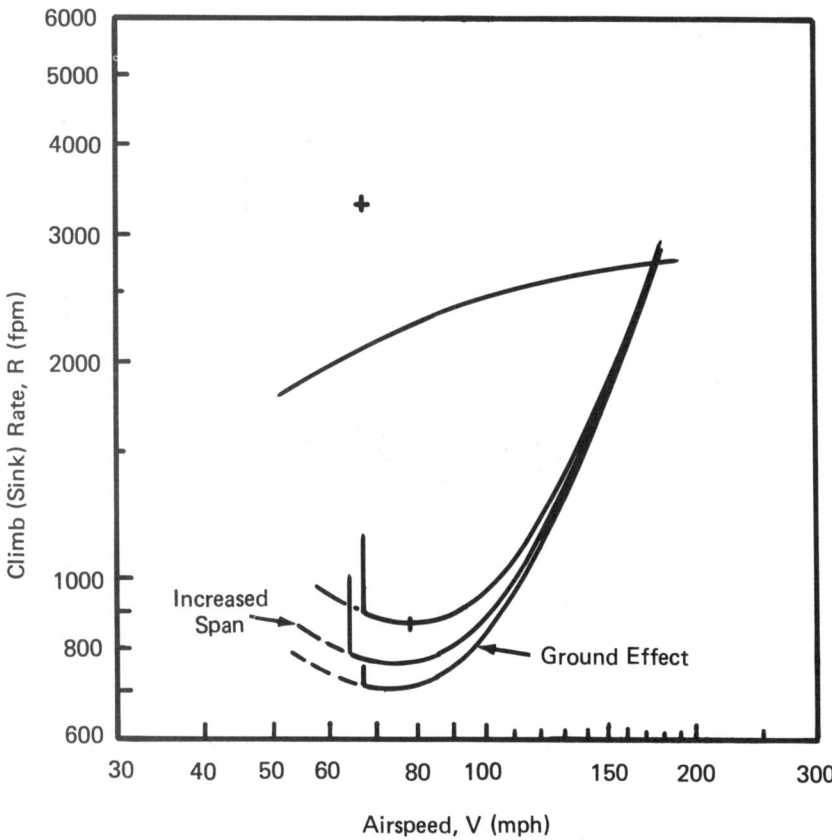

Figure 52. Effects of Increased Span and Ground Effect on Performance.

Next, we will see what the effect would be if we added another foot of span to the wing on each side of the fuselage. To separate this effect from all the others, we will assume that the wing span can be increased without increasing the total weight or drag of the airplane. Adding the extra amount of span without changing the chord, increases the total wing area and the aspect ratio. The increased wing area will result in a lower stall speed according to relation ①. The new span of 22 ft 10 inches will give a wing area of 94.3 ft^2, compared to the old wing area of 86 ft^2. This results in a new stall speed of 64 mph compared to 67 mph. The aspect ratio changes from 5.05 to the new value of 5.53. If we rework the calculation of the airplane efficiency factor as in Appendix F, we find that e is equal to 0.742, hardly changed from 0.744 from the previous analysis. Therefore, the new effective span is 20.83 $\sqrt{0.742}$ or 19.67 feet. This is 1.09 times larger (9 percent larger) than the effective span of our baseline design. In a manner similar to that discussed for the ground effect, we can draw a new curve for the sink rate as a function of the airspeed for the airplane with a 2 foot larger span. Note that the stall speed has decreased because of the added wing area. This follows from relation ①. The difference between the ideal rate-of-climb curve and the sink rate curve will give the climb rate as a function of the airspeed. This will be discussed after we have found the effects of propeller diameter on the performance in the next section.

POWER EFFECTS: LARGER ENGINE, TWIN VS. SINGLE ENGINE, PROPELLER DIAMETER

Let us consider three separate effects on the performance of the baseline design: increased power by adding a 180-hp engine, a theoretical redesign of the airplane so that we have a twin engined airplane with the same horsepower (two engines of 75 hp each); and a 10 percent increase in the propeller diameter. For these analyses we assume that all of the other parameters remain the same: drag area, weight, effective span, and maximum lift coefficient.

An increase in the horsepower from 150 to 180 is similar to the calculation previously made for power setting, except that now we have 1.2 times the original power. If we position the template for the ideal rate-of-climb so that we align the reference point with a BHP ratio of 1.2, we can draw the ideal climb rate as a function of airspeed as in Figure 53.

For our hypothetical twin, we have to be aware of what the reference point location actually does. We are trying to find the location of the point where $(V_{prop}, R^*_{C,max})$ is located. The ideal 74% efficiency speed, V_{prop}, is proportional to BHP$^{1/3}$. Therefore, since each engine has half of the total horsepower (and we have assumed that the propeller diameter has remained the same), the reference point moves to the left with a ratio $(1/2)^{1/3} = 0.794$, giving a value of 53 mph for V_{prop}. This results in the ideal rate of climb as given by the curve in Figure 53 for the twin-engined airplane. Also, while we are playing with the parameter study, we can find the performance with one engine out, since the airplane will have the same V_{prop} as for the twin, but half the ideal rate-of-climb. This is indicated in Figure 53.

The effect of increasing the propeller diameter is an increased propeller efficiency through the change in V_{prop}. This is the same effect that the twin makes. In fact, we would have to increase the propeller diameter by 41 percent to obtain the same benefits as making the airplane into a twin. But, to make a prop that size, we would have to slow down the propeller rotational speed and extend the landing gear. At any rate, the effect of the increased propeller size is shown in Figure 53. The increased propeller size can make quite a difference, however, if the design is marginal in the first place, as we will see later for the sample calculation of the performance of the powered Quicksilver hang glider.

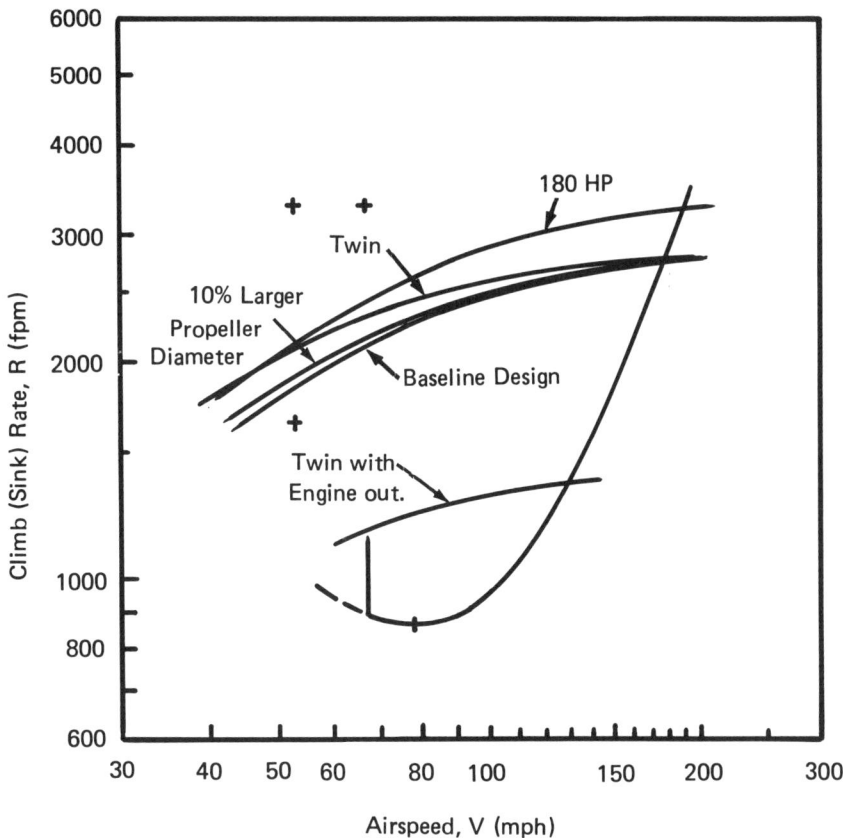

Figure 53. Effects of Engine Size, Twin vs. Single Engine, and Propeller Size on Performance.

COMPARISON OF THE PARAMETRIC VARIATIONS OF WEIGHT, DRAG, SPAN, AND POWER

The purpose of the parametric study is to determine the sensitivity of the design to variations in weight, drag, span, and power. Then, knowing the response to these changes, we can set priorities for improving our design. From Figures 50, 51, 52, and 53, we can subtract the sink rate from the ideal rate-of-climb to find the rate-of-climb as a function of the airspeed. All of the effects are plotted on Figure 54.

The effects that increase the climb rate the most are: the change from a 150 hp engine with the gross weight remaining the same; the decrease in weight by 150 lb (10%); conversion to a hypothetical twin-engined airplane with the same horsepower; ground effect at 5 feet altitude; an increase in span by 2 feet; a 10% larger diameter propeller; and the 10% decrease in the drag area. The largest decrease in climb rate was, of course, for the hypothetical twin with one engine out, while the climb rate for the overgross and flaps-extended cases were about the same. We find that the largest increase in level flight speed occurs for the increased engine horsepower. The next largest increase was for the case where the drag area was decreased by 10%. The changes in weight or effective span did not change the maximum speed significantly, because the effective span loading affects only the induced drag, which is small at high speeds. The maximum level flight speed is significantly decreased for the case where flaps are deployed and for the twin with the engine-out condition.

We have to make tradeoffs in the design of any airplane. For instance, we can decrease the drag area by using retractable landing gear, but the added mechanism will increase the weight. Therefore, we expect to have a higher maximum speed at the expense of a decrease in the rate-of-climb. Similarly, when we increase the engine horsepower of the airplane, we will probably have to increase the gross weight to have the same useful load. Tradeoffs between weight and drag will be governed by our desired specifications for rate-of-climb and maximum level speed.

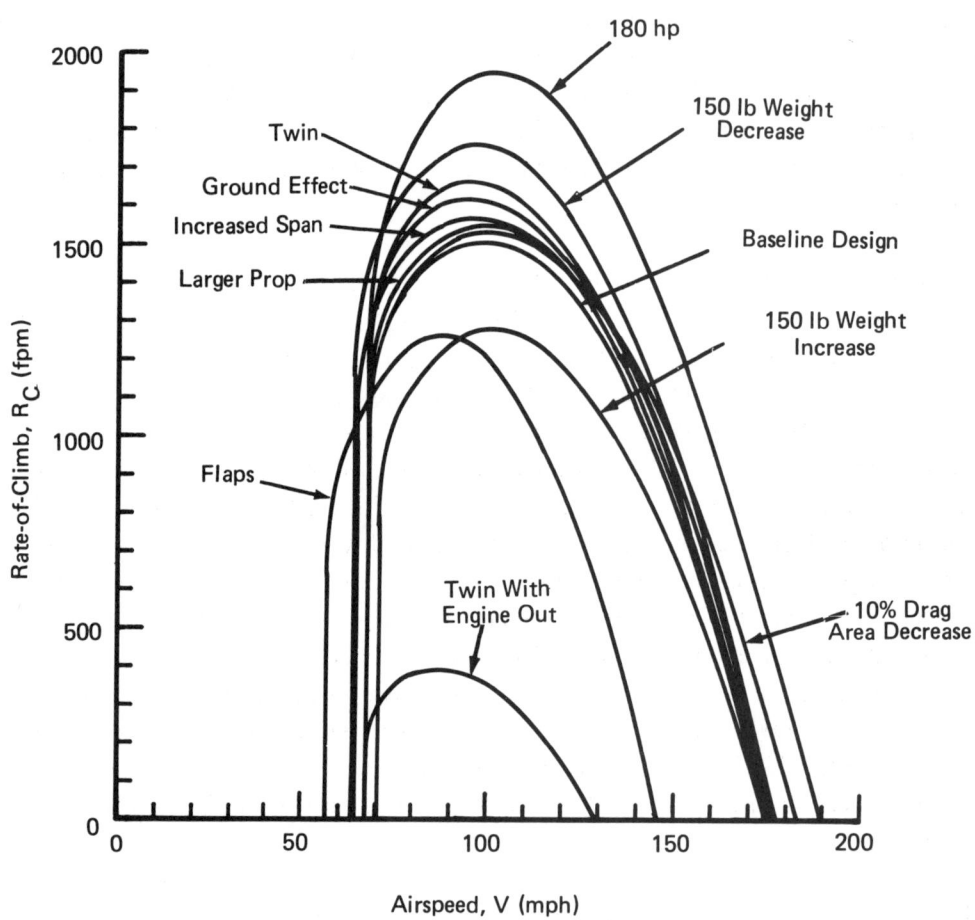

Figure 54. Parametric Effects of Weight, Drag Area, Power, Span, Propeller Diameter on Rate-of-Climb Performance.

SAMPLE CALCULATIONS OF AIRCRAFT PERFORMANCE

In this section we will show how to use the Airplane Performance and Design Nomogram and the template for ideal rate-of-climb and sink-rate to evaluate the performance of some known airplanes. Then, we will discuss the performance of the "Crawdad" — my original design study for a foot-launched motor-glider.

GOSSAMER CONDOR AND GOSSAMER ALBATROSS — MAN-POWERED AIRCRAFT

On August 23, 1977, at Shafter Airport in California, Dr. Paul MacCready's Gossamer Condor won the elusive Kremer Prize — £50,000 offered by the British industrialist for the successful man-powered flight around a figure-eight course embracing two turning points 1/2-mile apart with the start and finish at a height not less than ten feet above the ground. Since we know the weight and dimensions of the aircraft, and if we make an estimate of the drag area, we can estimate the performance by using the nomogram and template described in this book.

The gross weight of the aircraft is 207 lb (70 lb empty); the span is 96 ft; the average chord of the wing is 7.5 ft; the maximum available power from Bryan Allen, the pilot and engine, is about 0.48 hp; the propeller diameter is 12 ft 6 in; and an estimate of the drag area is about 23 ft^2. If the average height of the wing above the ground during flight is about 15 ft, the ratio of height to span is $h/b = 15/96 = 0.16$. With an analysis similar to that of Appendix F, we find that the airplane efficiency factor is 0.68 at higher altitudes and 1.03 in ground effect. The results of the calculation using the Airplane Performance and Design Nomogram are summarized on the nomogram itself in Figure 55. The maximum lift coefficient of 1.4 and a total wing area of 800 ft^2, including the lifting canard, were used for the stall analysis. The shape of the sink rate curve was modified in the vicinity of the stall speed, 8.5 mph, to give the curves in Figure 56 for the sink rate and the ideal climb rate. The calculation shows that the maximum level flight speed is 11.7 mph and the best rate of climb is about 12 fpm at 9-1/2 mph. If the template is used to find the minimum power input for level flight, we find that the power required is about 0.77 times the original maximum BHP (Bryan horsepower) or 0.37 hp at 9.5 mph. These numbers are in good agreement with the values given by MacCready (1978). The Gossamer Condor now hangs in a place of honor in the Smithsonian Air and Space Museum and a historical marker has been placed at Shafter Airport (near Bakersfield) commemorating the flight.

On June 12, 1979, Bryan Allen made worldwide news again when he pedaled the Gossamer Albatross across the English Channel. The craft is an improved version of the Gossamer Condor, with better streamlining and lighter weight to reduce the power requirements. The flight covered the 22 mile distance in 2 hours and 49 minutes, at altitudes less than 15 feet. The £100,000 prize was captured by Allen and MacCready, the designer, on the first try.

The performance calculations for the Gossamer Albatross can be considered as a parametric study of the original Gossamer Condor design with the major difference being the reduction in the drag area from 23 ft^2 to 11 ft^2. The wing had the same span, but was tapered and had a smaller average chord. The empty weight was reduced to 55 lb, but the pilot was required to carry a life jacket as a safety precaution. Consequently, the gross weight was about the same as for the Condor. The drag area ratio is about 0.48 (= 11/23), which can be used to position the template to obtain the new sink rate curve in Figure 56. The

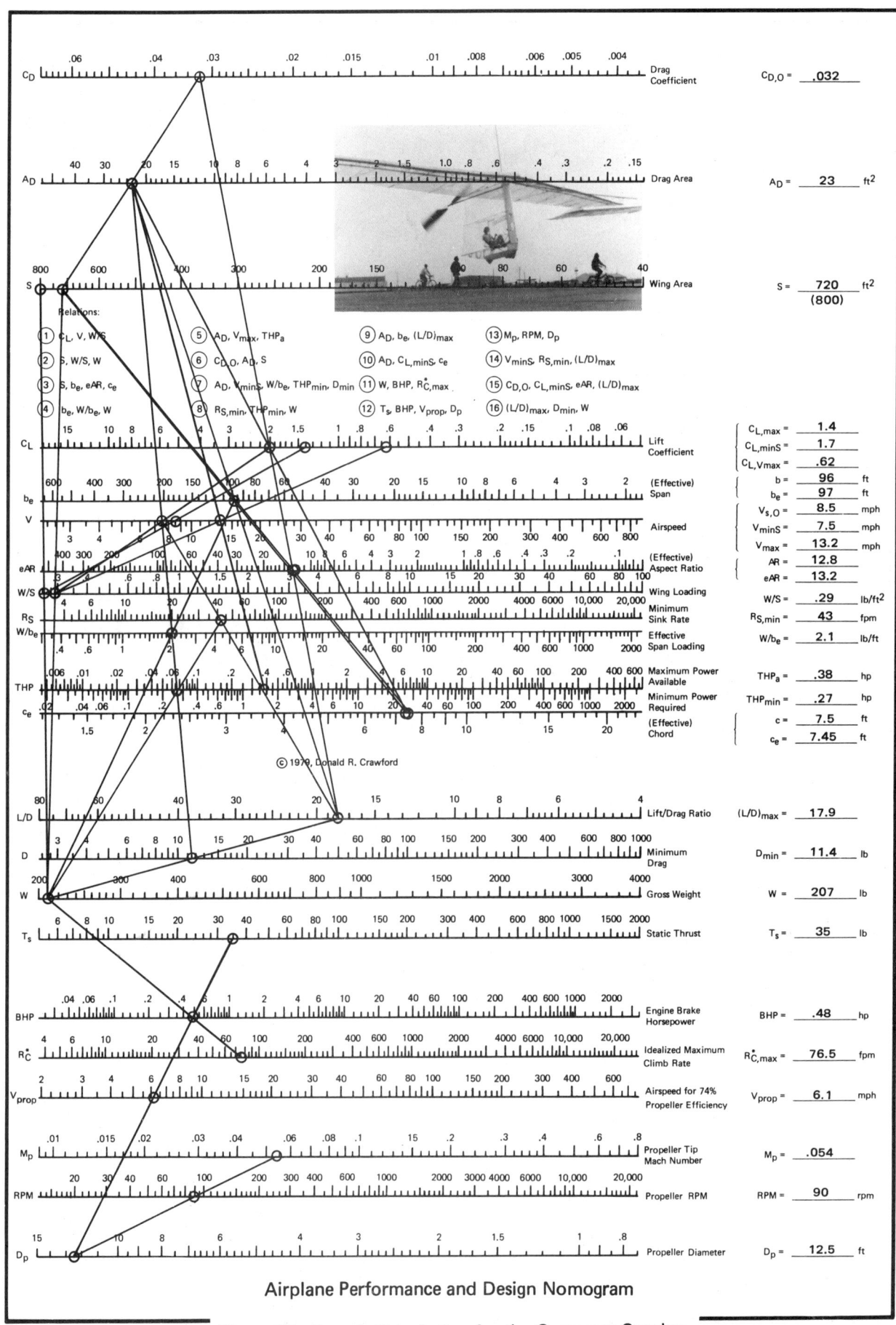

Figure 55. Sample Calculation for the Gossamer Condor.

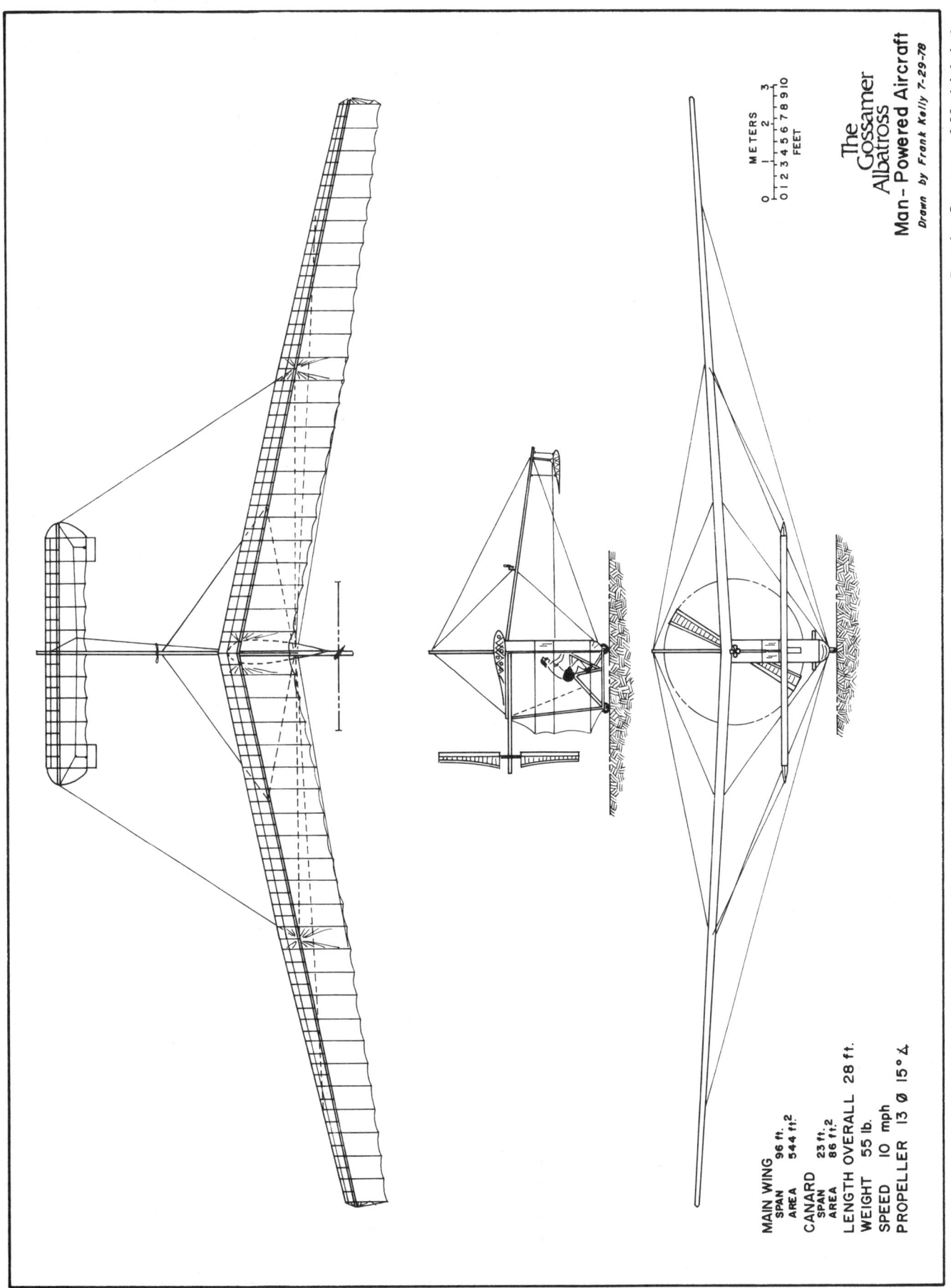

minimum power required for level flight can be found by sliding the climb rate template with the reference point aligned with the BHP ratio line until the sink rate and the climb rate just touch. The BHP ratio turns out to be 0.59 for minimum power. Multiplying by 0.48 HP, we find that the minimum power required for level flight of the Gossamer Albatross is about 0.28 HP. The maximum level flight speed is found to be about 15.8 mph.

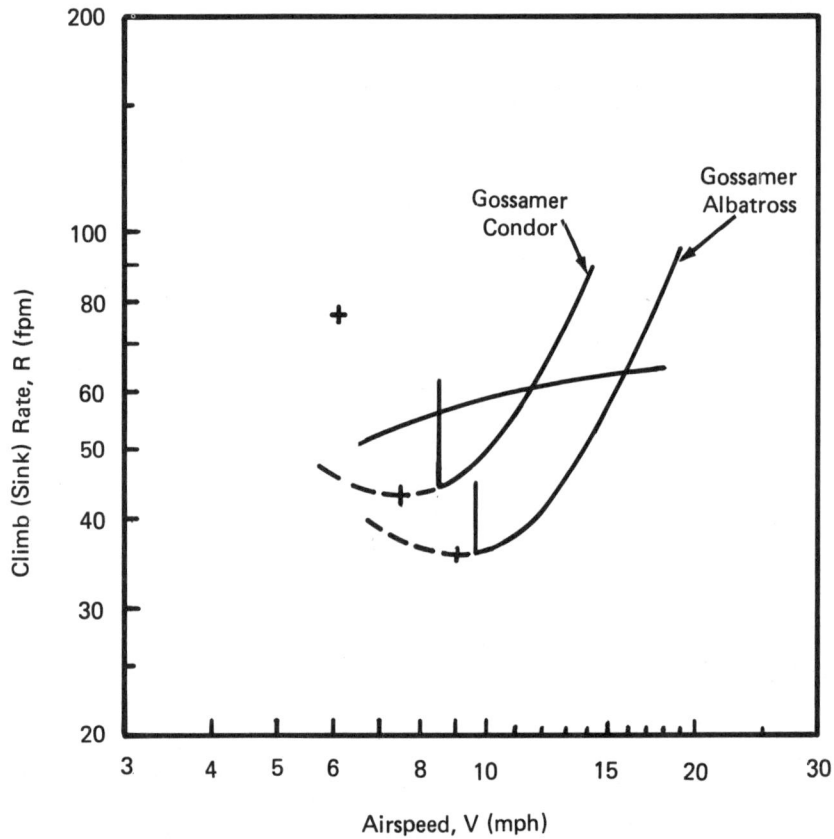

Figure 56. Performance of Gossamer Condor and Gossamer Albatross.

POWERED QUICKSILVER HANG GLIDER --- MICROLIGHT

This sample calculation is aimed at trying to understand how the performance of a powered hang glider is improved by the addition of a gear reduction unit and a slower-turning, larger-diameter propeller. The use of the small go-kart engines which turn at high (7000) rpm limit the size of the propeller to keep the tip speeds small enough to obtain good efficiency. If we limit the tip Mach number to 0.8, the 7000 rpm will give a maximum prop diameter of 29-1/2 inches. The propeller diameter used on Bob Bowen's Quicksilver was 28 inches when it was used as a direct drive propeller.

If we take the following parameters as known input data, we can calculate the performance from the Airplane Performance and Design Nomogram and the template for ideal climb rate and sink rate vs airspeed: wingspan 30 ft; chord 5 ft; the empty weight 75 lb and the gross weight 243 lb; the power of the MAC 101 12 hp; the propeller diameter 28 inches and the assumed drag area of about 16 ft^2 (6 ft^2 for the seated pilot and 10 ft^2 for the wires, tubes and dacron wings). If the maximum lift coefficient is assumed to be

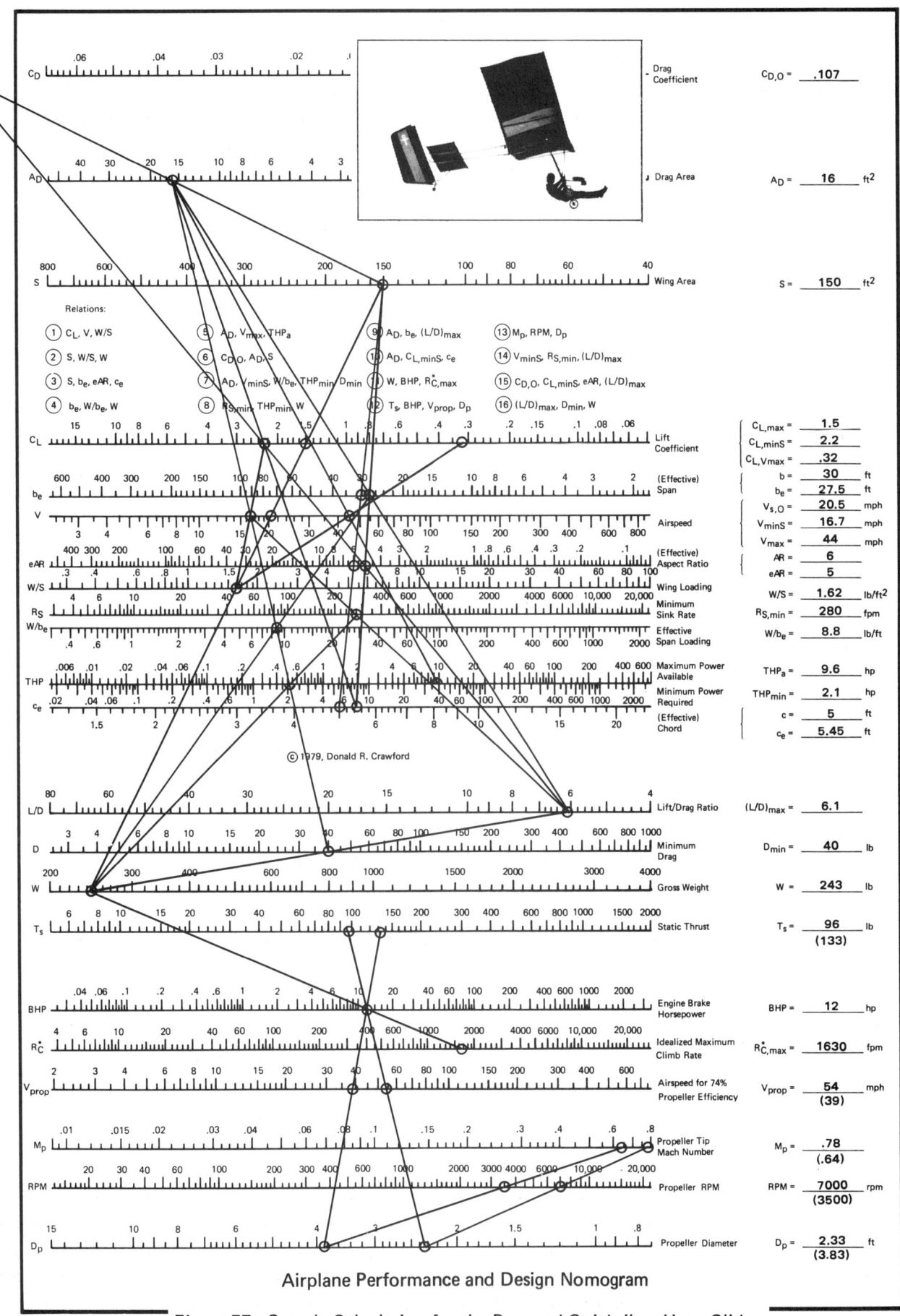

Figure 57. Sample Calculation for the Powered Quicksilver Hang Glider.

1.5, the stall speed is 20.5 mph. This is larger than his quoted landing speed (15 mph) but this can probably be accounted for by recognizing the unsteady motion involved in the landing technique, where there is a bird-like landing flare. With the aspect ratio of 6 we can use the analysis of Appendix F to estimate the airplane efficiency factor. This value was taken to be 0.84 for the wing only, neglecting the effect of the fuselage, since the pilot hanging out in the air does not contribute any angle-of-attack dependent drag variation. The effective span was therefore taken to be 27½ ft. These values were used on the nomogram in Figure 57, with the remaining parameters tabulated on the nomogram in the blank spots provided.

If the propeller is replaced by a larger diameter propeller (46 inches) and a 2:1 gear reduction unit so that the propeller turns at 3500 rpm, the reference propeller speed, V_{prop}, decreases substantially, resulting in a more efficient propeller at each speed (if the pitch chosen is optimum). The ideal rate-of-climb and sink rate are plotted in Figure 58. The resulting climb rates for the two cases are plotted in Figure 59. It can be seen that the new geared propeller gives far better climb performance compared to the old direct drive propeller.

If the machine is limited to 30 mph cruise speeds (for pilot comfort) we find that only half throttle is required with the new prop combination. With the direct drive unit, the throttle setting would theoretically be about 67%. These types of calculations are made by sliding the template so that the reference point for the ideal climb rate stays on the Brake Horsepower Ratio scale until the climb-rate curve crosses the sink-rate curve. (See the sample calculation for the power setting of the T-18.)

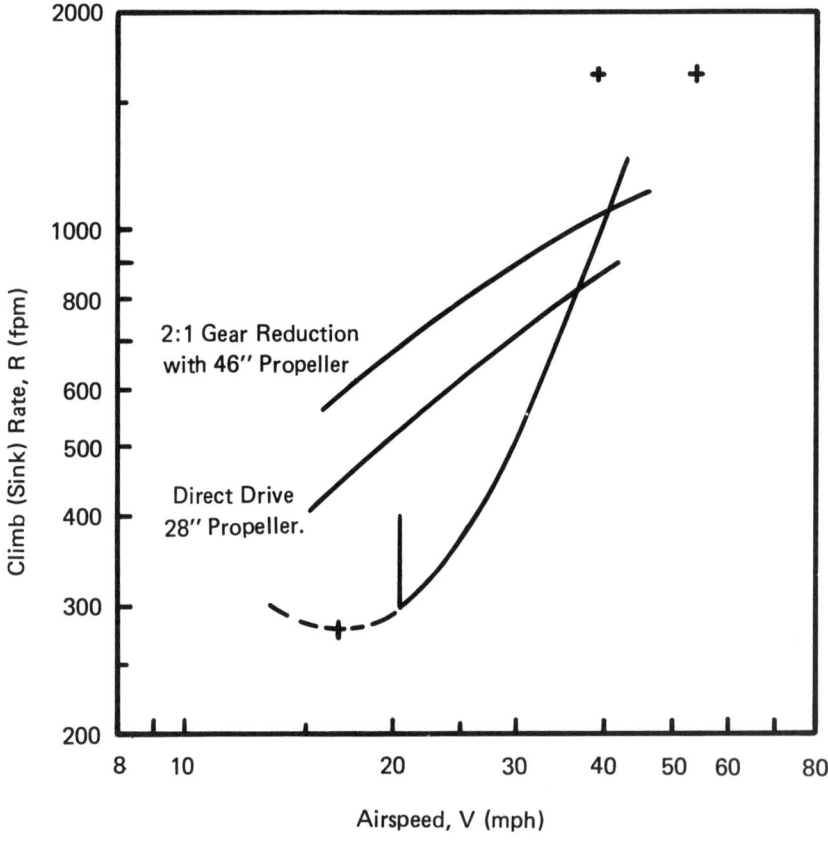

Figure 58. Effects of Propeller Size on Performance of a Powered Quicksilver Hang Glider.

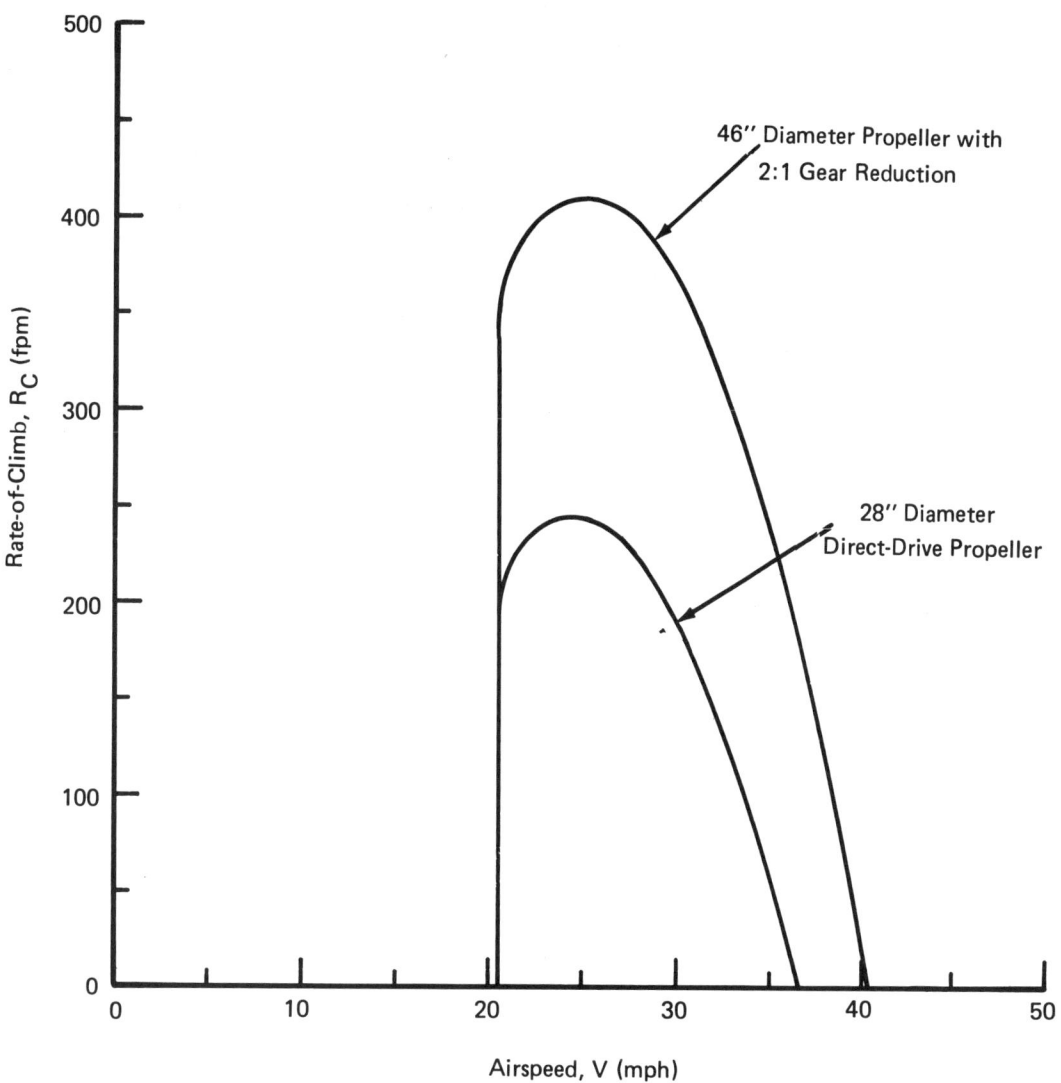

Figure 59. Effect of Propeller Diameter on Powered Quicksilver Hang Glider.

CESSNA 172 — GENERAL AVIATION AIRCRAFT

The Cessna 172 Skyhawk has been the most popular 4-place general aviation aircraft for more than ten years. The 172 was introduced in 1956, and has been steadily improved year by year. The performance calculations were made assuming the following information: wing span 35 ft 10 in; wing area 174 ft^2; empty weight 1379 lb; gross weight 2300 lb; powerplant 160-hp Lycoming 0-320-H; propeller diameter 6 ft 3 in; maximum lift coefficeint 1.6 (NACA 2412); estimated zero-lift drag coefficient 0.032 [Hoerner (1965) for Cessna 170]. The remaining performance parameters are summarized on the Airplane Performance and Design Nomogram in Figure 60. The airplane efficiency factor was taken to be 0.77 following an analysis similar to that in Appendix F for an airplane with a round fuselage and rectangular wings.

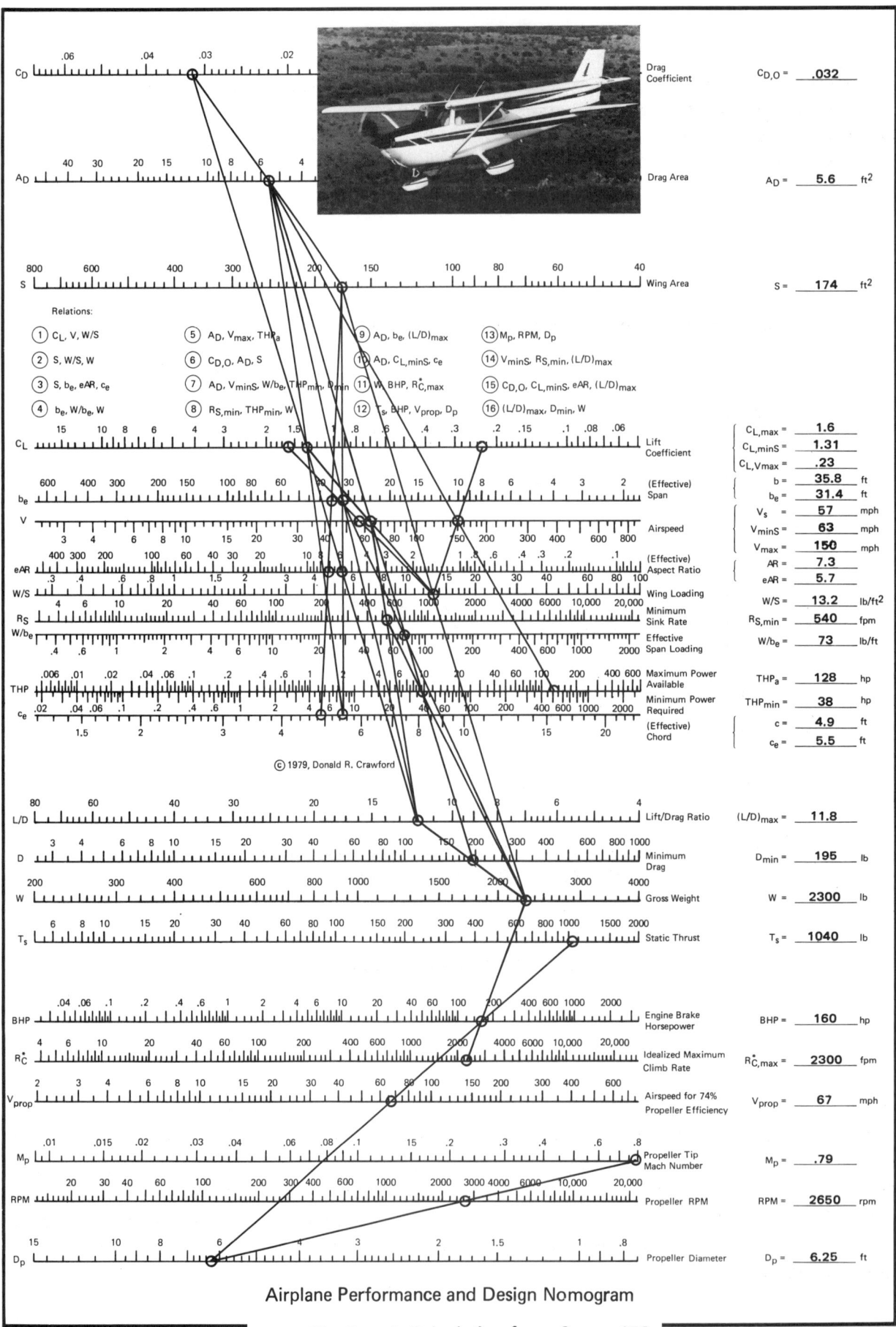

Figure 60. Sample Calculation for a Cessna 172.

With the lift coefficient and the wing loading, we find that the flaps up stall speed is 57 mph, which agrees with the value from the operating manual. When the sink rate and ideal climb rate are used to position the template, we can draw the curves in Figure 61. The difference between the curves gives the rate-of-climb as plotted in Figure 62.

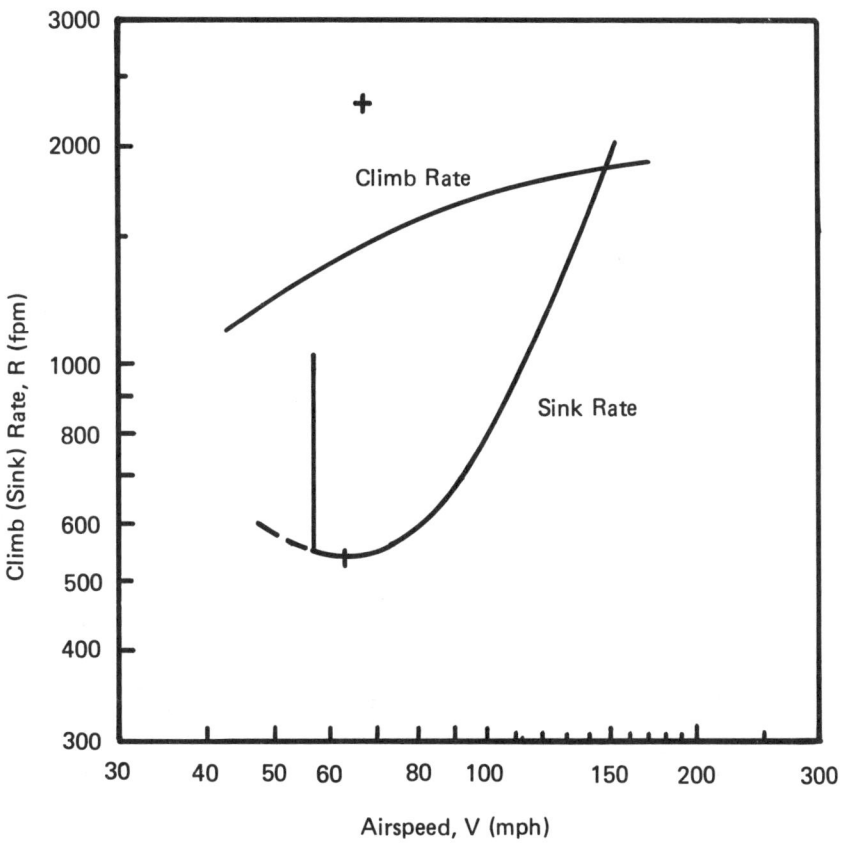

Figure 61. Performance of Cessna 172.

The rate-of-climb curve appears to be overly optimistic, with a maximum of almost 1000 ft/min, compared to the listed 770 ft/min. The maximum level flight speed is calculated at 147 mph, while the data listed in **Jane's** is 144 mph. Although these numbers are not perfect, they give a good idea of the expected performance for the airplane. One explanation might be that the engine with the fixed pitch propeller is not able to attain the optimum rpm to put out the maximum available horsepower. The brake horsepower actually available would have to be accounted for if more accuracy is desired for the performance calculations.

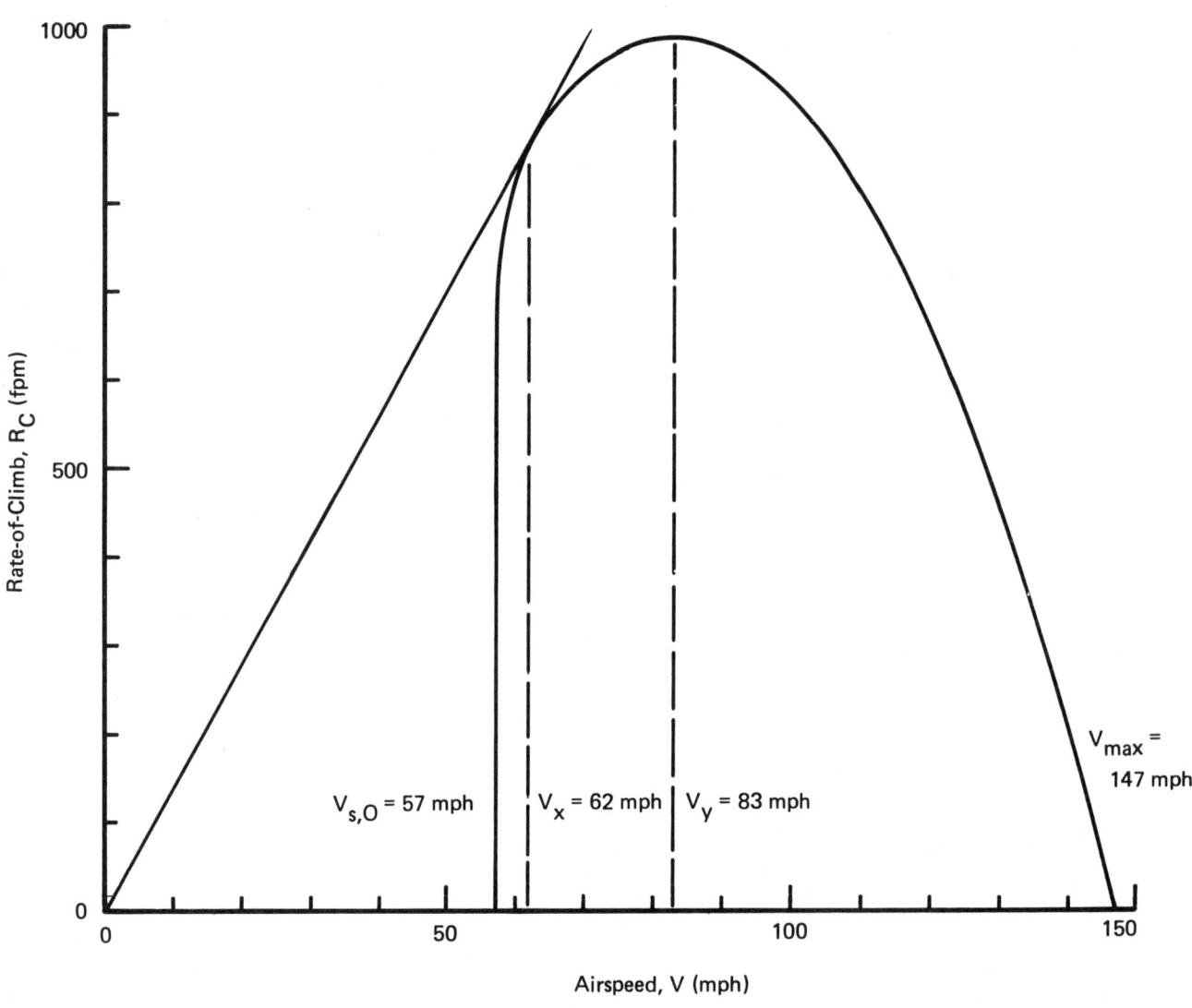

Figure 62. Rate-of-Climb Performance for Cessna 172.

"CRAWDAD" — FOOT-LAUNCHED MOTORGLIDER DESIGN STUDY

This discussion is the result of a design study for a foot-launched motorglider (Figure 63) that I hope to build in the near future. The main constraint on the design is that it should be capable of being foot-launched without assistance using the engine I obtained last year at the EAA Fly-In at Chino — the Air Craft Marine Engineering Co. 30-hp two cylinder two-stroke engine. At the recommended speed of 4500 rpm the engine develops 22.4 hp. The following design parameters were assumed for the performance calculations: empty weight 144 lb; maximum gross weight 330 lb; stall speed without flaps 20 mph; maximum lift coefficient without flaps 1.4; maximum power 22.4 hp; propeller diameter 44 inches; wingspan 37 ft; maximum lift coefficient with flaps 2.1; estimated drag area 7.1 ft^2. With an analysis similar to that in Appendix F, we find that the airplane efficiency factor is about 0.81.

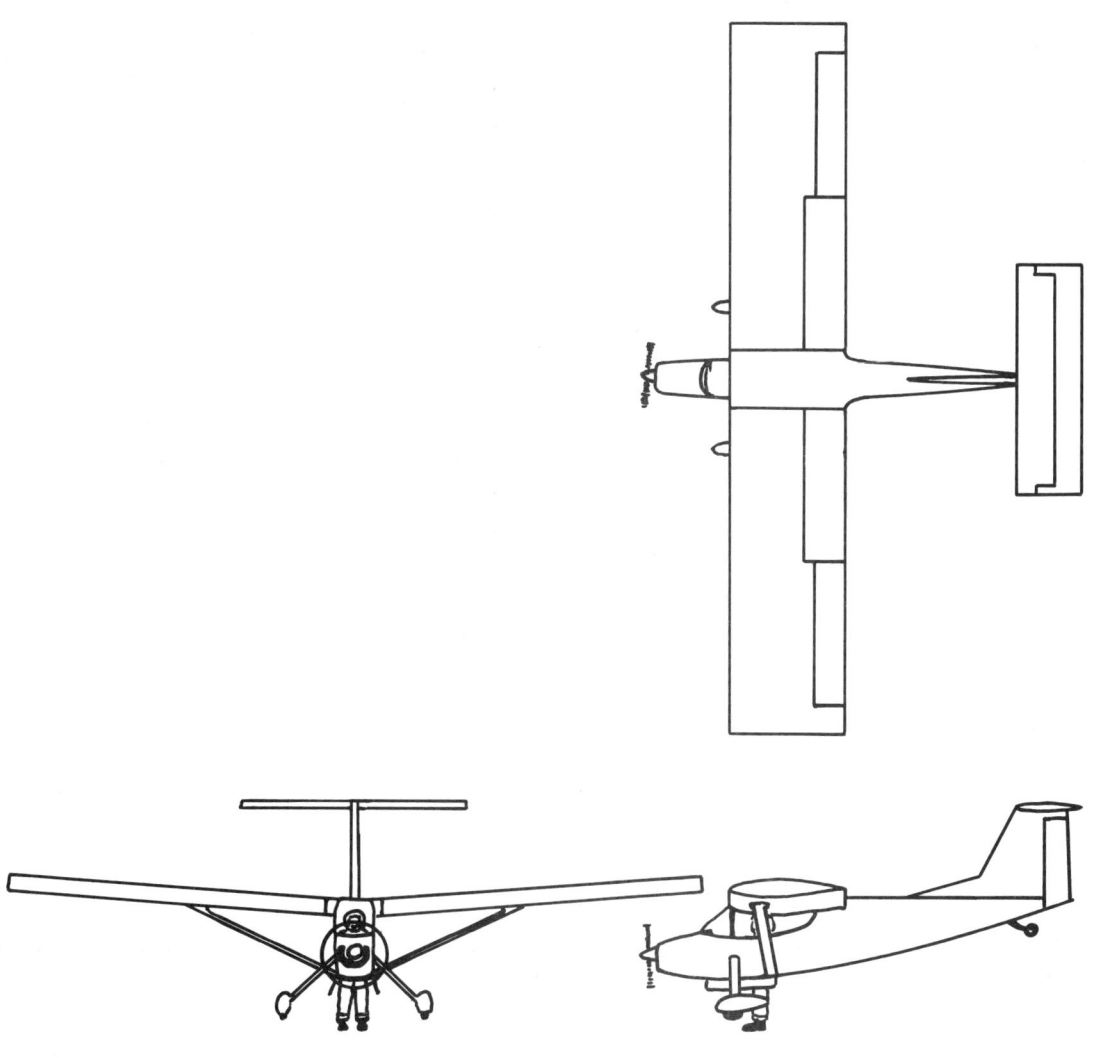

Figure 63. Preliminary 3-View of Crawdad — A Foot-Launched Motorglider.

With the assumed design parameters we can construct the nomogram in Figure 64 and tabulate the remaining variables in the blank spaces provided. The results are then used to construct the ideal rate-of-climb and sink-rate curves in Figure 65. The performance data is summarized in Table 1. The three-view seems to suggest that the weight may be larger than the 144 lb budgeted. This will be better defined with a more careful set of weight and balance calculations to determine the empty weight and the center of gravity of the airplane at maximum takeoff weight conditions. A mockup of the cockpit will be made before further details of the design are pursued.

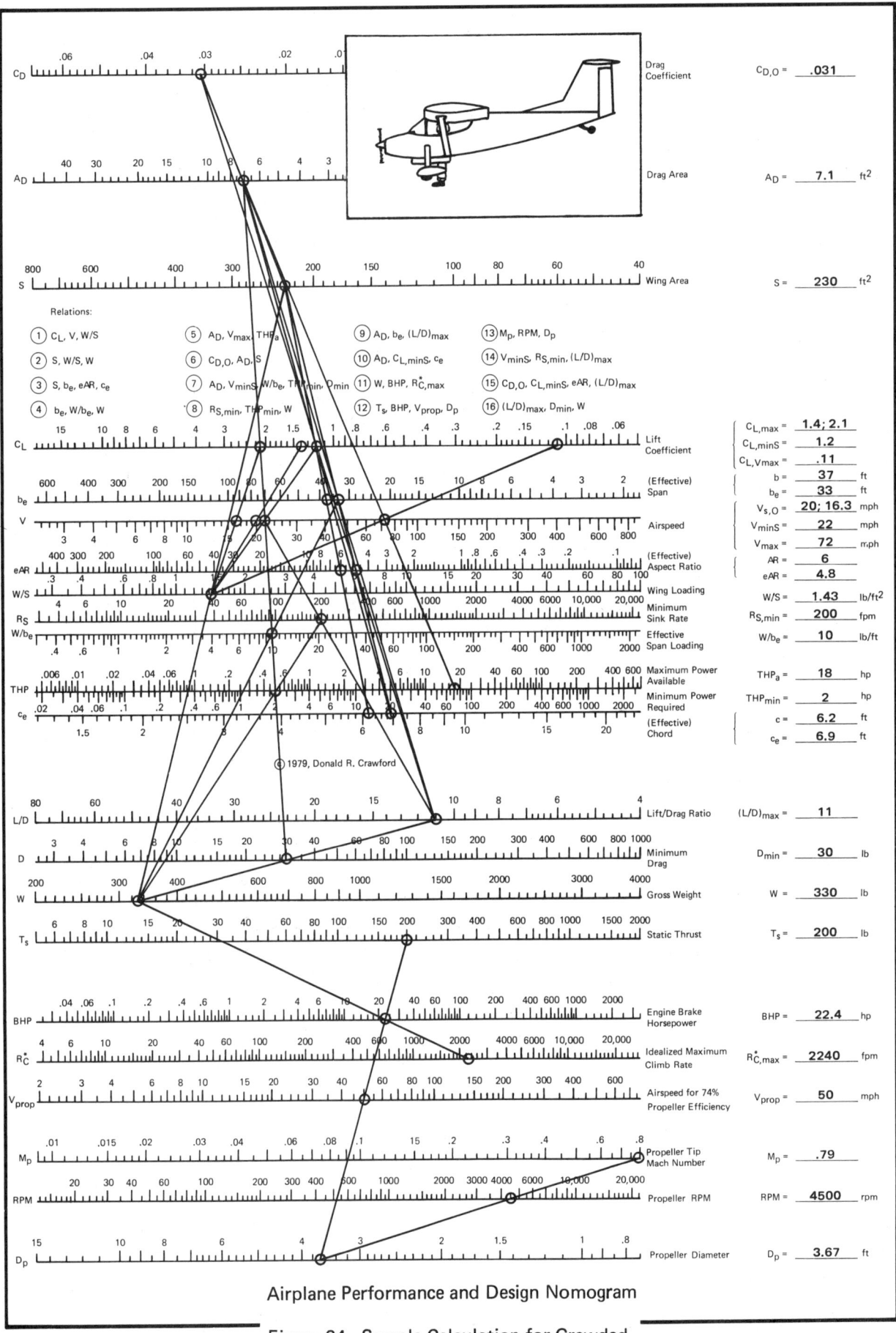

Figure 64. Sample Calculation for Crawdad.

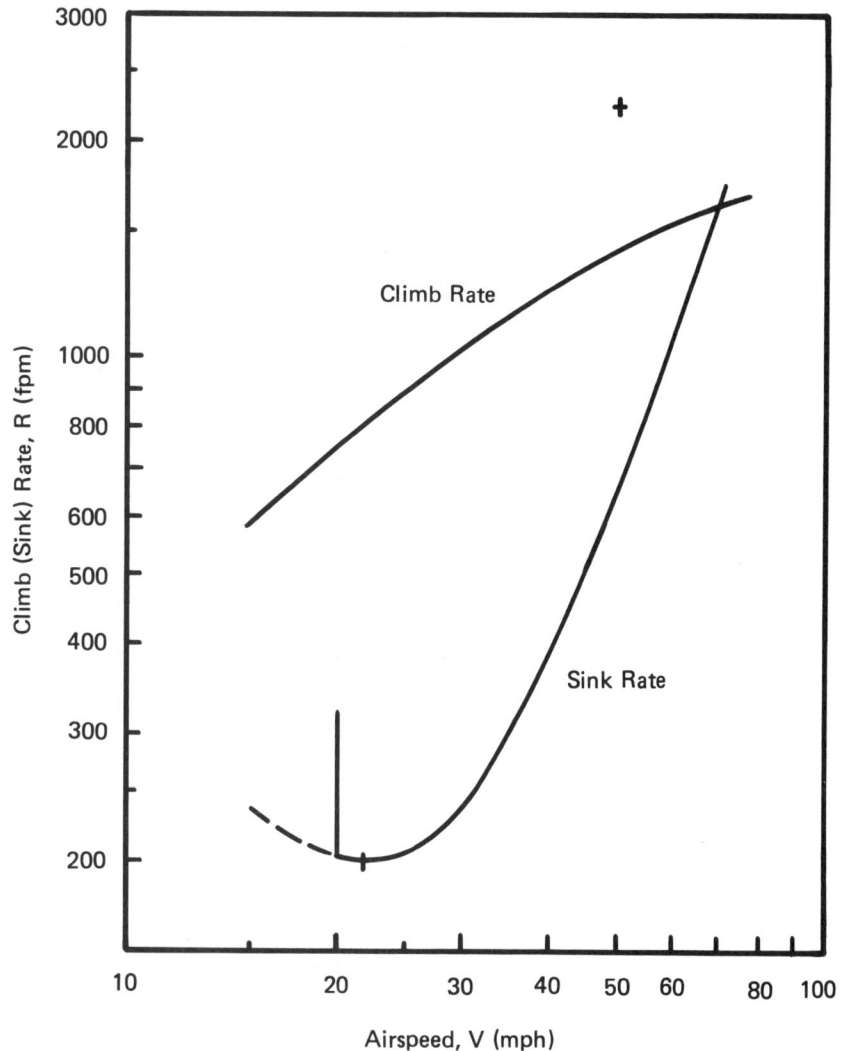

Figure 65. Performance of Crawdad.

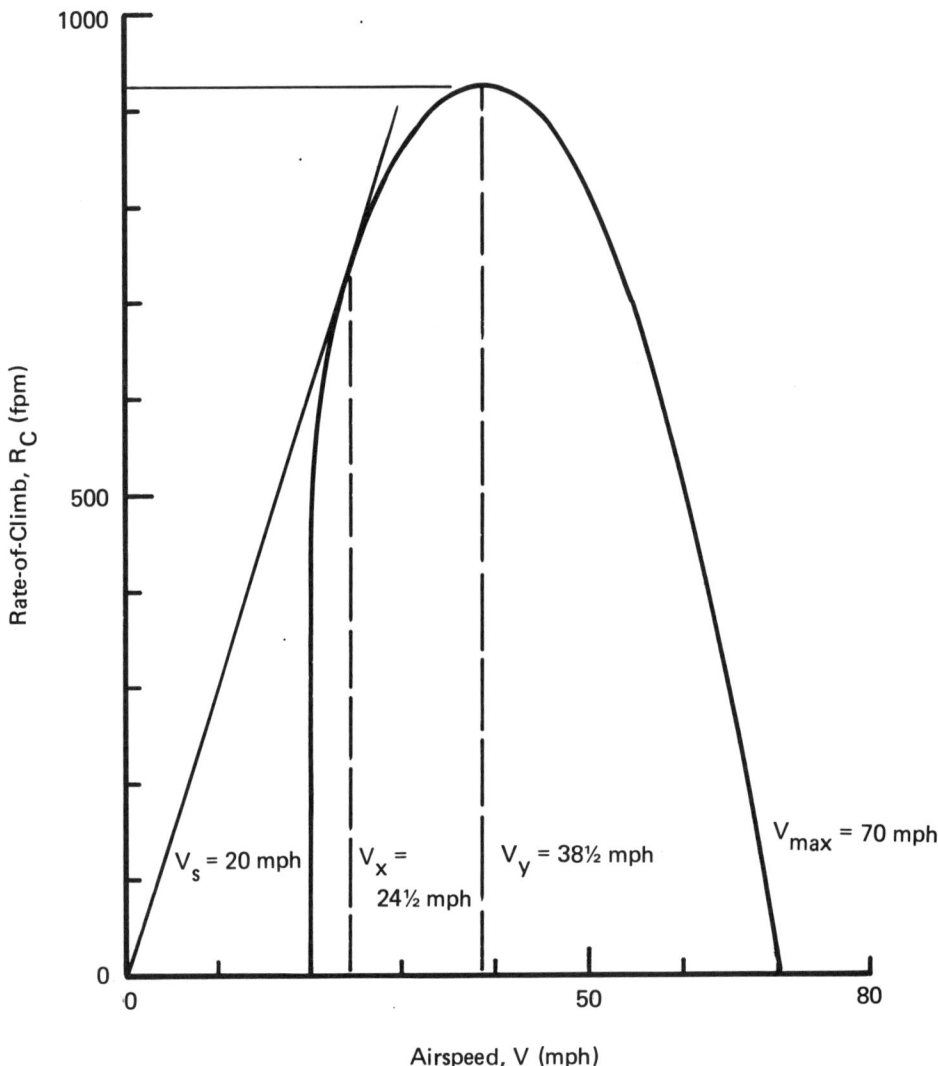

Figure 66. Calculated Rate-of-Climb Performance for "Crawdad".

Table 1. Summary of Preliminary Performance Parameters for the Sample Calculation of "Crawdad" — a Foot-Launched Motor Glider

*Lift Coefficient at Stall (without Flaps)	$C_{L,max1} = 1.4$
*Stall Speed without Flaps	$V_{s,1} = 20$ mph
Wing Loading	$W/S = 1.43$ lb/ft^2
*Lift Coefficient with Flaps	$C_{L,max0} = 2.1$
Stall Speed with Flaps	$V_{s,O} = 16.3$ mph
Weight of the Powerplant	$W_{pp} = 45$ lb
Weight of the Structure	$W_s = 99$ lb
Empty Weight	$W_e = 144$ lb
Weight of Payload (Pilot)	$W_p = 180$ lb
Weight of the Fuel (1 gallon)	$W_f = 6$ lb
Useful Load	$W_u = 186$ lb
*Gross Weight	$W = 330$ lb
Wing Area	$S = 230$ ft^2
*Wing Span	$b = 37$ ft
Aspect Ratio	$AR = 6.$
Wing Chord	$c = 6.23$ ft = [75 in]
†Aircraft Efficiency Factor	$e = .81$
Effective Aspect Ratio	$eAR = 4.8$
Effective Span	$b_e = 33.3$ ft
Effective Chord	$c_e = 6.91$ ft

*Assumed Value
†Estimated Value Using Appendix F
‡Estimated Value Using Appendix G

Table 1. (Cont.)

Effective Span Loading	$W/b_e = 9.91$ lb/ft
‡Drag Area	$A_D = 7.1$ ft^2
Zero-Lift Drag Coefficient	$C_{D,O} = .031$
Speed for Minimum Sink Rate	$V_{minS} = 21.8$ mph
Minimum Power Required for Level Flight	$THP_{min} = 2.0$ hp
Minimum Drag Force	$D_{min} = 29.8$ lb
Maximum Lift-to-Drag Ratio	$(L/D)_{max} = 11.1$
Minimum Sink Rate	$R_{S,min} = 200$ fpm
Lift Coefficient at Minimum Sink Power Conditions	$C_{L,minS} = 1.18$
Engine Brake Horsepower (@ 4500 RPM)	$BHP = 22.4$ hp
Ideal Rate of Climb	$R^*_{C,max} = 2240$ fpm
*Propeller Diameter	$D_p = 3.67$ ft = 44 in
Characteristic Propeller Velocity	$V_{prop} = 50$ mph
Idealized Static Thrust	$T_s = 180$ lb
Propeller Tip Mach Number	$M_p = 0.79$
Maximum Cruise Speed	$V_{max} = 70$ mph
Maximum Rate of Climb at Sea Level	$R_{C,max} = 925$ fpm
Speed for Best Rate of Climb	$V_y = 38.5$ mph
Speed for Best Angle of Climb	$V_x = 24.5$ mph
Performance Rating Parameter	$F_p = \dfrac{(R/C)_{max} W_u}{33000\, BHP} \left[1 - \dfrac{V_{s,O}}{V_{max}}\right] = 0.167$
Kinetic Energy Parameter	$WV^2_{max} = 1.64 \times 10^6$ lb mph^2

AIRPLANE PERFORMANCE COMPARISONS

It is sometimes difficult to quantify the performance of one design with respect to another. Is top speed the only criterion, or should the weight carrying capacity and the rate of climb also be considered? In this section we will reintroduce a "performance rating parameter," F_p, that can be used to compare performance efficiency of various aircraft. In addition, a measure of the size of an aircraft is needed so that we can separate the various classes of aircraft — microlight, ultralight, etc.

The performance rating parameter was suggested by Ross (1948) and is a combination of three efficiency ratios: (1) useful load to maximum gross weight, W_u/W; (2) excess available thrust horsepower to maximum available brake horsepower, $\Delta THP_a/BHP$; and (3) speed range (maximum speed minus minimum speed) to maximum speed, $(V_{max} - V_{min})/V_{max}$. Therefore, if we had a perfect airplane, it would have all of the useful load equal to the maximum gross weight, could put all of the available power into climb, and could fly like a helicopter. The range of F_p will always be between zero and unity. Since the maximum excess horsepower can be related to the maximum weight and the maximum rate of climb, $R_{C,max}$, we can redefine the performance rating parameter by

$$F_p = \frac{W_u R_{C,max}}{33,000 \, BHP} \left[1 - \frac{V_{min}}{V_{max}} \right] < 1$$

Since this relation holds equally for ordinary airplanes as well as for helicopters, we can compare various airplanes by substituting the performance data into the above relation. There is no need to measure the wing area or the gross weight. If we try to increase our performance rating by increasing the useful load, we will probably suffer by the reduction in the rate-of-climb. The operation of an efficiency contest, such as the Pazmany Efficiency Contest held each year at the EAA Convention at Oshkosh, might be harder to judge because of the difficulty of measuring rate-of-climb, but the results would be a better measure of the overall aircraft efficiency.

The performance rating parameter, F_p, has been calculated for a large number of aircraft from the performance data obtained from **Jane's All the Worlds Aircraft** as listed in Appendix L. The performance rating parameter is plotted against the kinetic energy parameter (described below) in Figure 67. We can calculate the values for our airplane or proposed design and see how the data point compares to those of other aircraft. The performance ratings of the aircraft shown on the figure can be considered as goals to surpass in our future designs.

It should be emphasized that the numbers used are based on data usually supplied by the manufacturer and may not have been verified by test. Care is needed when making direct comparisons of the airplane performance rating parameter.

The other variable that might be of interest is a kinetic energy parameter WV_{max}^2. This variable is a measure of the destructive energy that an airplane would have at its maximum flying speed, and can be considered as a measure of its overall size. If the size of the kinetic energy parameter is small enough, the airplane has little destructive potential so that there should be no need for regulation by the Federal Aviation Administration. Powered hang gliders could fit into a class such as this without having to be restricted to those that are only capable of foot-launching — as long as they cannot fly at too large a maximum speed. The kinetic energy parameter may be useful when comparing a small fast airplane to a larger, but slower, machine.

For the sample calculation of the Thorp T-18, the performance rating parameter and the kinetic energy parameter can be computed from the performance estimates. The useful load is 606 lb; the maximum rate of climb is 1500 fpm; the brake horsepower is 150 hp; the stall speed is 67 mph; the maximum level flight speed is 175 mph; and the gross weight is 1500 lb. Therefore, the performance rating parameter is

$$F_p = \frac{606 \quad 1500}{33000 \quad 150} \left[1 - \frac{67}{175} \right] = 0.113$$

The kinetic energy parameter is

$$WV_{max}^2 = 1500 \quad 175^2 = 4.59 \times 10^7 \text{ lb mph}^2$$

This information is plotted on Figure 67.

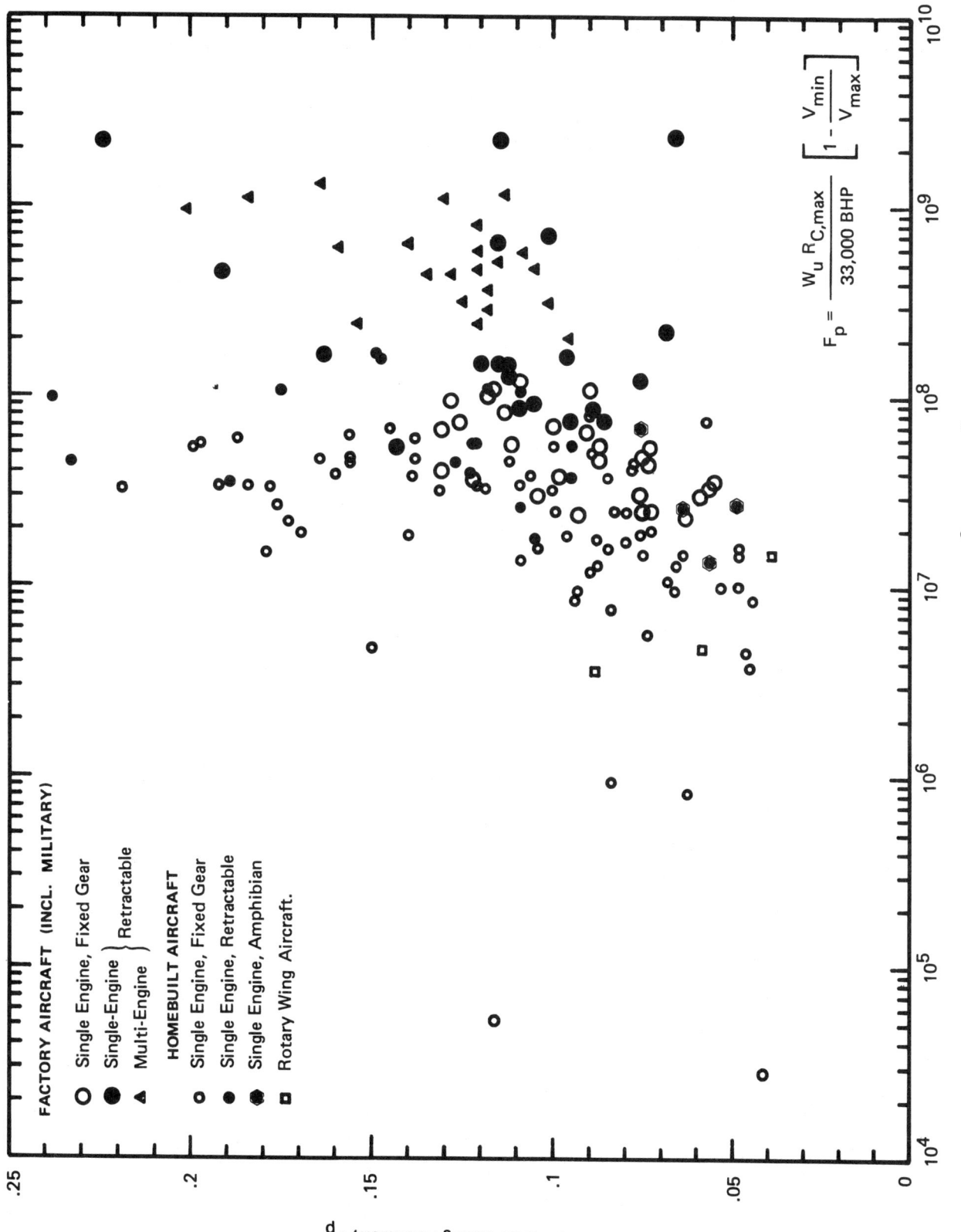

Figure 67. Performance Rating Comparison.

PART 2. THEORETICAL BACKGROUND

LOW SPEED AERODYNAMICS

The equations governing the flight of an airplane in equilibrium can be broken down into two groups: those for gliding and powered flight. In gliding flight, we will find the sink rate as a function of the relative airspeed, with the altitude and the airplane weight, effective span, and drag area as parameters. The sink rate times the weight (equal to the drag times the relative airspeed) gives precisely the power required to maintain level flight at that airspeed, when care is taken with the dimensions of force, velocity, and power. Therefore, when the thrust-power that is available from the engine-propeller combination is equal to this value, the airplane can maintain level flight. If additional power is added, the airplane can climb while at the same airspeed, or can fly faster in level flight. The power available from the thrust of the propeller will be determined from a simplified analysis. The propeller is replaced by an idealized disk which instantaneously accelerates the air to produce thrust to propel the airplane. If it is assumed that the engine is used to lift a weight equal to that of the airplane, then the rate at which the weight is lifted minus the sink rate gives the actual rate of climb for the airplane. This result will hold when it is assumed that the glide and climb angles are small (less than 20 degrees, say) and that the thrust acts in the same direction as the freestream velocity vector.

FORCE BALANCE IN GLIDING FLIGHT

Let us consider the force balance in equilibrium gliding flight as shown in Figure 68. It is assumed that the wing and tail moments are neglected, and that the freestream density is constant (although the density can vary slowly with altitude). Then, a force balance in the s-direction (along the flight path) can be written as

$$W \sin \theta_g = D \tag{1}$$

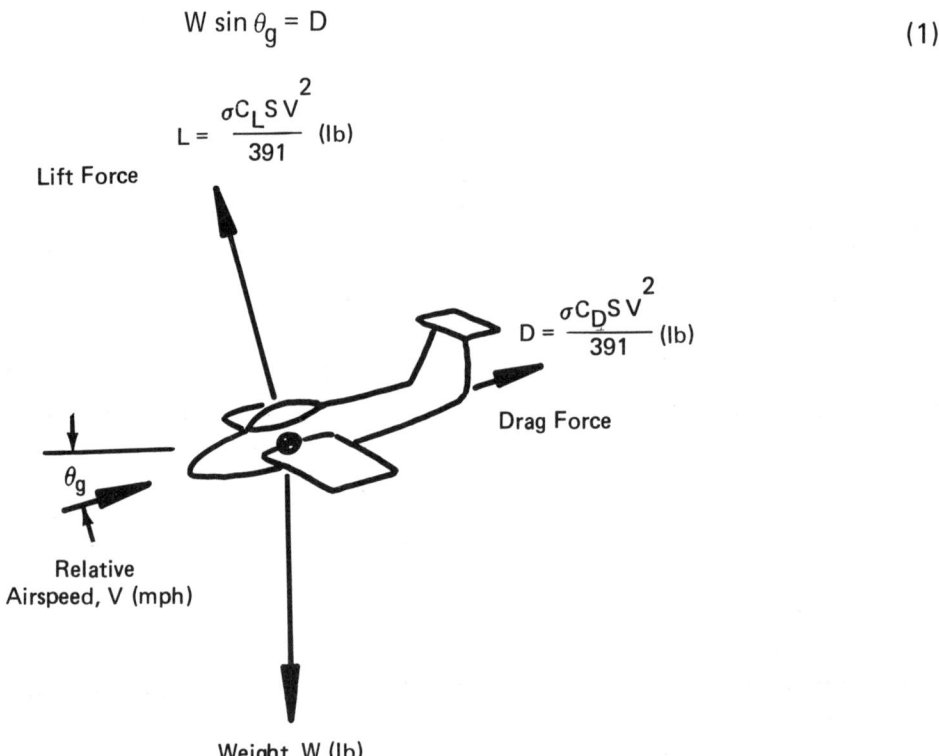

Figure 68. Force Balance for Gliding Flight in Equilibrium.

where W is the airplane weight (lb), D is the drag force acting on the airplane (lb) and θ_g is the glide angle (degrees). The drag force is assumed to include the parasite drag and the drag due to lift — the induced drag. A similar force balance in the n-direction (normal to the flight path) gives

$$W \cos \theta_g = L \qquad (2)$$

where L is the lift force of the wing. If the forces were not in equilibrium, we would also need to consider the acceleration forces. Since the lift and drag forces have the same dimensions as a pressure times an area, we can define the dimensionless lift and drag coefficients, C_L and C_D, by

$$C_L = \frac{L}{\frac{1}{2} \rho V^2 S} \qquad (3)$$

and

$$C_D = \frac{D}{\frac{1}{2} \rho V^2 S} \qquad (4)$$

The term $\frac{1}{2}\rho V^2$ is called the dynamic pressure (lb/ft^2), where ρ is the air density (slugs/ft^3), V is the relative airspeed (ft/sec), and S is the reference area (ft^2) — usually taken to be the wing area if C_L and C_D are the lift and drag coefficients for the airplane. If we express the velocity in miles per hour, introduce the density ratio σ (the ratio of the freestream density, ρ, to standard sea-level density, $\rho_{SL} = 0.002377$ slugs/ft^3), and use (3) and (4), then Equations (1) and (2) become

$$W \sin \theta_g = \frac{\sigma C_D S V^2}{391} \qquad (5)$$

and

$$W \cos \theta_g = \frac{\sigma C_L S V^2}{391} \qquad (6)$$

The altitude dependence of σ is discussed in Appendix D and plotted in Figure 43.

If we make the small angle approximation, the cosine of the glide angle is approximately unity, and the sine of the glide angle is approximately equal to $[\theta_g \pi /180]$. Therefore, equation (6) becomes

$$\frac{W}{S} = \frac{\sigma C_L V^2}{391} \qquad (7)$$

which is discussed in Part 1 as relation ①. The term W/S is defined as wing loading — the weight divided by the wing area (relation ②). The glide angle, θ_g, is found from equation (5) using the small angle approximation

$$\theta_g = \frac{180}{\pi} \frac{\sigma C_D S V^2}{W \, 391} \qquad (8)$$

If the velocity from (7) is substituted into (8), we have

$$\theta_g = \frac{180}{\pi} \frac{C_D}{C_L} \qquad (9)$$

so that the glide angle depends only upon the ratio of the drag and lift coefficients. The glide angle does not depend upon the altitude (or density ratio, σ) or upon the weight.

If we define the sink rate, R_S, by $V \sin \theta_g$, then

$$R_S = V \frac{\sigma C_D S V^2}{391 W} \times 88 \quad \text{(ft/min)}, \qquad (10)$$

where the factor 88 converts miles per hour into feet per minute. Now, if the airspeed is determined from equation (7) and substituted into (10), we have

$$R_S = 88 \sqrt{\frac{391 W}{\sigma S}} \frac{C_D}{C_L^{3/2}} \quad \text{(ft/min)}. \qquad (11)$$

We see, therefore, that for minimum sink rate, we need to minimize the term $C_D/C_L^{3/2}$; for minimum glide angle, we need to minimize C_D/C_L. We will now discuss the dependence of the drag coefficient on the lift coefficient for finite span wings.

INDUCED DRAG

For a finite span airplane we can split the drag coefficient into two parts, one for parasite drag, $C_{D,O}$ and one for induced drag, $C_{D,i}$.

$$C_D = C_{D,O} + C_{D,i} \tag{12}$$

The parasite drag of the airplane includes the profile drag of the airfoil alone, as well as the skin friction drag, pressure drag and interference drag on the other aircraft components. If the wing has an elliptic lift distribution, as shown in Figure 69, Prandtl's lifting line theory relates the induced drag coefficient to the lift coefficient by

$$C_{D,i} = \frac{C_L^2}{\pi AR} \tag{13}$$

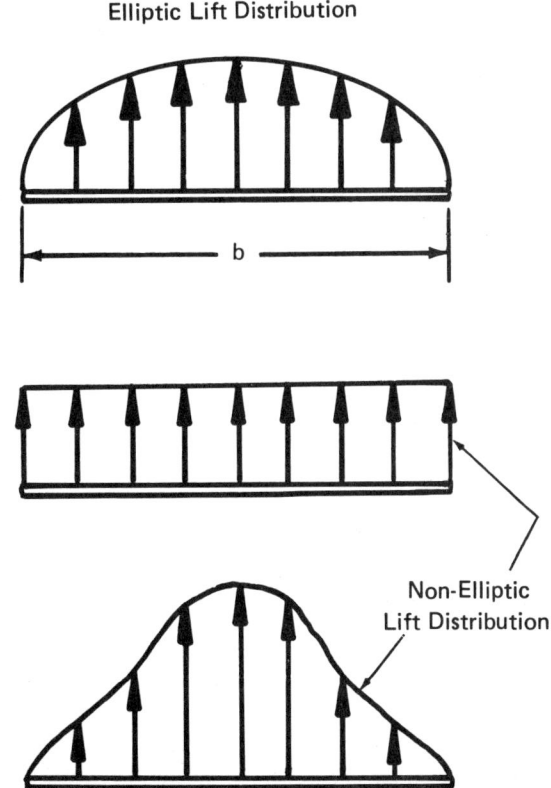

Figure 69. Spanwise Distribution of Lift on the Wing.

where AR is the aspect ratio of the wing. The aspect ratio is related to the span, b, wing area, S, and average chord, c, by the relation

$$AR = \frac{b^2}{S} = \frac{b}{c}, \quad (14)$$

which is relation ③ on the nomogram. However, if the wing does not have an elliptic lift distribution, the induced drag contribution is multiplied by an amount 1/e, where e is the airplane efficiency factor and is usually less than 1.0. The efficiency factor, e, depends on the planform shape of the wing, airfoil section, wing twist, wing sweepback angle, fuselage cross-section shape, and ground effect (when the airplane is flying less than one wing span above the ground).

When the airplane efficiency (also called the Oswald efficiency factor after a noted aerodynamicist of the 30's) is taken into account, the airplane drag coefficient can be written as

$$C_D = C_{D,O} + \frac{C_L^2}{\pi e AR}. \quad (15)$$

This is the definition of the so-called parabolic drag polar where the curve for C_D as a function of C_L has the shape of a parabola. Departures from a parabolic form for the lift-drag polar would require a modified analysis such as that given by Wendt (1947). The combination eAR is called the effective aspect ratio since a wing with this aspect ratio would follow Prantdl's theory. The practical way in which this value is determined is discussed in Appendix F.

MINIMUM SINK RATE

The minimum sink rate can be determined by substituting the drag polar equation (15) into (11) and using calculus to find the value of C_L that minimizes $C_D/C_L^{3/2}$. The term $C_D/C_L^{3/2}$ is given by

$$\frac{C_D}{C_L^{3/2}} = \frac{C_{D,O}}{C_L^{3/2}} + \frac{C_L^{1/2}}{\pi e AR} . \qquad (16)$$

If we differentiate this expression with respect to C_L and set the result equal to zero, we have

$$\frac{d}{dC_L}\left[\frac{C_D}{C_L^{3/2}}\right] = -\frac{3}{2}\frac{C_{D,O}}{C_L^{5/2}} + \frac{1}{2}\frac{1}{\pi e AR}\frac{1}{C_L^{1/2}} = 0 . \qquad (17)$$

Solving for C_L, we find that the value for C_L that minimizes $C_D/C_L^{3/2}$ is

$$C_{L,minS} = \sqrt{3\pi e AR \, C_{D,O}} . \qquad (18)$$

On the nomogram, this expression is part of relation ⑮.

Let us introduce the drag area, A_D. This is the area of a plate placed normal to the air flow that would produce the same zero-lift drag as the complete airplane. If the drag coefficient of the plate is equal to 1.0, then equating the drag would give $A_D = C_{D,O} S$, (which is relation ⑥). If we solve for $C_{D,O}$ from this definition, substitute into (18), use equation (14) to eliminate AR and define the effective chord by $c_e \doteq c/\sqrt{e}$, we find that the lift coefficient that minimizes the sink rate is

$$C_{L,minS} = \sqrt{3\pi}\,\frac{\sqrt{A_D}}{c_e} \qquad (19)$$

This is relation ⑩.

From (18) and (19) we see that there is a tradeoff between effective aspect ratio and zero-lift drag coefficient on one hand, and drag area and effective chord on the other. Since we would like to keep the lift coefficient at minimum sink conditions below the maximum lift coefficient value, we need to have a well streamlined airplane (keep $C_{D,O}$ small) if we have a large effective aspect ratio. This is the case for sailplanes. On the other hand, if we have a large drag area — such as for hang gliders where the pilot and flying wires are hanging out in the air — we need to have a large effective chord. If either of these criteria is not met, the minimum sink condition will occur very near stall and it becomes dangerous to try to maintain the absolute minimum sink condition while soaring. When the lift coefficient for minimum sink turns out to be greater than the maximum lift coefficient, the theory breaks down and we have to make changes in the stall region. A practical method for this modification is discussed in Part 1 in the section for sink rate versus airspeed (R_S vs V).

If the lift coefficient that was calculated to minimum $C_D/C_L^{3/2}$ is substituted into (15), we find that the induced drag is three times the parasite drag, resulting in a drag coefficient four times the zero-lift drag coefficient. If these results are substituted into (11), we find that the minimum sink rate is given by

$$R_{S,min} = \underbrace{88 \sqrt{391} \frac{4}{(3\pi)^{3/4}}}_{1294} \sqrt{\frac{W}{\sigma}} \frac{A_D^{1/4}}{b_e^{3/2}} \quad \text{(ft/min)}, \qquad (20)$$

where the effective span, $b_e = b\sqrt{e}$, has been introduced. The sink rate is strongly dependent on effective span (-3/2 power), and less dependent on the weight (1/2 power), density ratio (-1/2 power), and drag area (1/4 power).

The airspeed at which the minimum sink rate occurs is found by substituting (19) into (7) and solving for V,

$$V_{minS} = \underbrace{\frac{\sqrt{391}}{(3\pi)^{1/4}}}_{11.285} \frac{\sqrt{W/b_e}}{\sqrt{\sigma} \, A_D^{1/4}} \quad \text{(mph)} . \qquad (21)$$

We see that effective span loading, W/b_e (relation ④), is a natural parameter that enters into the expression for the speed at minimum sink conditions. Equation (21) is represented on the nomogram as relation ⑦.

An alternative expression for the sink rate can be found if we substitute (15) and (7) into (11) and use the definitions for drag area and effective span,

$$R_S = 88 \left[\frac{\sigma A_D V^3}{391 \, W} + \frac{391 \, W}{\pi \sigma V b_e^2} \right] \quad \text{(ft/min)} . \qquad (22)$$

The first term represents sink rate due to the parasite drag, while the second term is due to induced drag. The minimum sink rate can also be found by differentiating this expression with respect to the airspeed and setting the result equal to zero.

$$\frac{dR_S}{dV} = 88 \left[\frac{3\sigma A_D V^2}{391 W} - \frac{391 W}{\pi \sigma V^2 b_e^2} \right] = 0 \quad . \tag{23}$$

Solving for V, we find that the minimum sink condition occurs when

$$V_{minS} = \frac{\sqrt{391}}{(3\pi)^{1/4}} \frac{\sqrt{W/b_e}}{\sqrt{\sigma} \, A_D^{1/4}} \quad \text{(mph)} \quad . \tag{24}$$

which is the same as (21).

If we divide (22) by (20), and define the dimensionless sink rate, $\widehat{R}_S = R_S/R_{S,min}$, and the dimensionless airspeed, $\widehat{V} = V/V_{minS}$, we obtain the equation for the dimensionless sink rate in terms of the dimensionless airspeed

$$\widehat{R}_S = \frac{\widehat{V}^3}{4} + \frac{3}{4\widehat{V}} \quad . \tag{25}$$

This equation is plotted in Figure 37 and is the equation that forms the basis for the construction of the sink rate versus airspeed curves using the plastic template included with the text. The manner in which the template is used in described in Part 1.

MAXIMUM LIFT-TO-DRAG RATIO

The lift-to-drag ratio is related to the glide angle through equation (9), since the ratio of the lift-force to the drag-force is equal to the ratio of the lift coefficient to the drag coefficient. To minimize the glide angle, we have to maximize the lift-to-drag ratio. First, substitute (15) into (9) and differentiate with respect to C_L and set equal to zero.

$$\frac{d\theta_g}{dC_L} = \frac{180}{\pi} \frac{d}{dC_L}\left[\frac{C_{D,O}}{C_L} + \frac{C_L}{\pi eAR}\right] = \frac{180}{\pi}\left[-\frac{C_{D,O}}{C_L^2} + \frac{1}{\pi eAR}\right] = 0 \quad (26)$$

Solving for C_L, we find the lift coefficient that minimizes the glide angle and maximizes the lift-to-drag ratio is

$$C_{L,maxLD} = \sqrt{\pi eAR\, C_{D,O}} = \frac{C_{L,minS}}{\sqrt{3}} \quad (27)$$

We see that the lift coefficient for maximum L/D is smaller than the lift coefficient for minimum sink. (If $C_{L,minS} = 1.$, then $C_{L,maxLD} = 0.577$.) In this case, the induced drag is equal to the parasite drag and the maximum lift-to-drag ratio is

$$(L/D)_{max} = \frac{\sqrt{\pi eAR\, C_{D,O}}}{2\, C_{D,O}} = \underbrace{\frac{\sqrt{\pi}}{2}}_{0.8862}\sqrt{\frac{eAR}{C_{D,O}}} \quad (28)$$

This expression is contained within relation ⑮ on the nomogram. If we introduce the drag area and the effective span, we can rewrite (29) as

$$(L/D)_{max} = \frac{\sqrt{\pi}\, b_e}{2\sqrt{A_D}} = 0.8862\, \frac{b_e}{\sqrt{A_D}} \quad (29)$$

We see that the best glide ratio is obtained for a large effective span and a small drag area, or equivalently, for a large effective aspect ratio and a small drag coefficient.

The airspeed for best L/D is found by substituting (27) into (7). Since the velocity ratio V_{maxLD}/V_{minS} is equal to $\sqrt{C_{L,minS}/C_{L,maxLD}}$, the speed for best L/D is $3^{1/4}$ times that for minimum sink, or about 32 percent larger. Since the lift is equal to the weight in the small angle approximation, the drag is a minimum when the glide angle is the smallest. In this case, the drag is equal to

$$D_{min} = \frac{2}{\sqrt{\pi}}\sqrt{A_D}\,\frac{W}{b_e} = 1.128\sqrt{A_D}\,\frac{W}{b_e} \quad (30)$$

Again, we see the importance of the effective span loading, W/b_e, and the drag area, A_D. This expression is part of relation ⑦ on the nomogram.

LEVEL FLIGHT

The force balance for equilibrium level flight is shown in Figure 70. Now, the drag force is balanced by the thrust produced by the propeller. Since the available thrust horsepower is equal to the thrust times the airspeed (with an appropriate conversion factor to change into units of horsepower), we have

$$THP_{a,L} = \frac{88}{33000} \left[\frac{\sigma A_D V^3}{391} + \frac{391 (W/b_e)^2}{\pi \sigma V} \right] \qquad (31)$$

Comparing this expression to (22), we see that the thrust horsepower required to maintain level flight is equal to the gliding sink rate that would occur at the same airspeed times the weight divided by 33000 — the factor that changes foot-pounds per minute into horsepower.

$$THP_{a,L} = \frac{R_S W}{33000} \qquad (32)$$

(Relation ⑧) the minimum power required for level flight therefore occurs at the same speed that minimizes the sink rate, V_{minS}. The minimum power required is

$$THP_{min} = \underbrace{\frac{88}{33000} \frac{\sqrt[4]{391}}{(3\pi)^{3/4}}}_{0.03921} \frac{A_D^{1/4}}{\sqrt{\sigma}} \left[\frac{W}{b_e} \right]^{3/2} \qquad (33)$$

This expression is part of relation ⑦

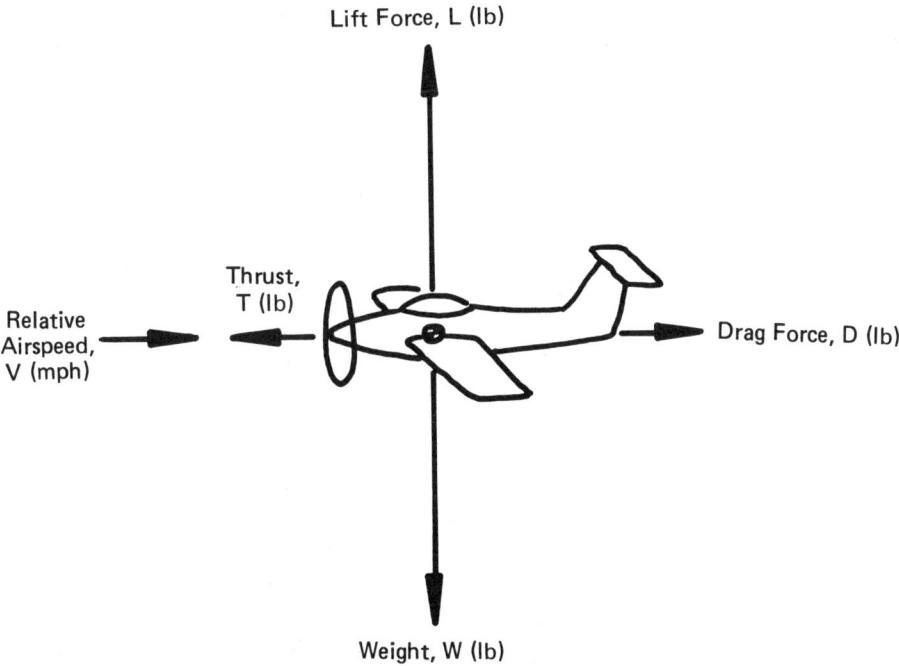

Figure 70. Force Balance for Level Flight in Equilibrium.

CLIMBING FLIGHT

The equilibrium force balance for climbing flight is shown in Figure 71. In the s-direction along the flight path, we have

$$T = D + W \sin \theta_c \tag{34}$$

and in the n-direction, normal to the flight path, we have

$$L = W \cos \theta_c \,, \tag{35}$$

where θ_c is the climb angle. If it is assumed that the climb angle is small, equation (35) reduces to (7) and equation (34) becomes

$$T = W \sin \theta_c + \frac{\sigma A_D V^2}{391} + \frac{391 \, (W/b_e)^2}{\sigma V^2} \,. \tag{36}$$

If we multiply by the relative airspeed, we find that the thrust-horsepower available to climb at this angle is

$$THP_a = \frac{W \, R_C}{33,000} + THP_{a,L} \,. \tag{37}$$

That is, the thrust horsepower available to climb at a rate R_C (ft/min) is equal to the weight times the climb rate divided by the conversion factor, plus the power required to maintain level flight. We can turn this relation around, since the power available is equal to the power required, and find that the rate of climb is

$$R_C = \left[\frac{33,000 \, BHP}{W} \right] \eta - R_S \,, \tag{38}$$

where the efficiency, $\eta = THP_a/BHP$, has been introduced. The term in brackets in equation (38) is found in relation ⑪ on the nomogram and is denoted by the symbol $R^*_{C,max}$, the idealized maximum rate-of-climb that would result if the weight of the airplane were lifted by the engine brake horsepower. The propeller efficiency is discussed in the next section.

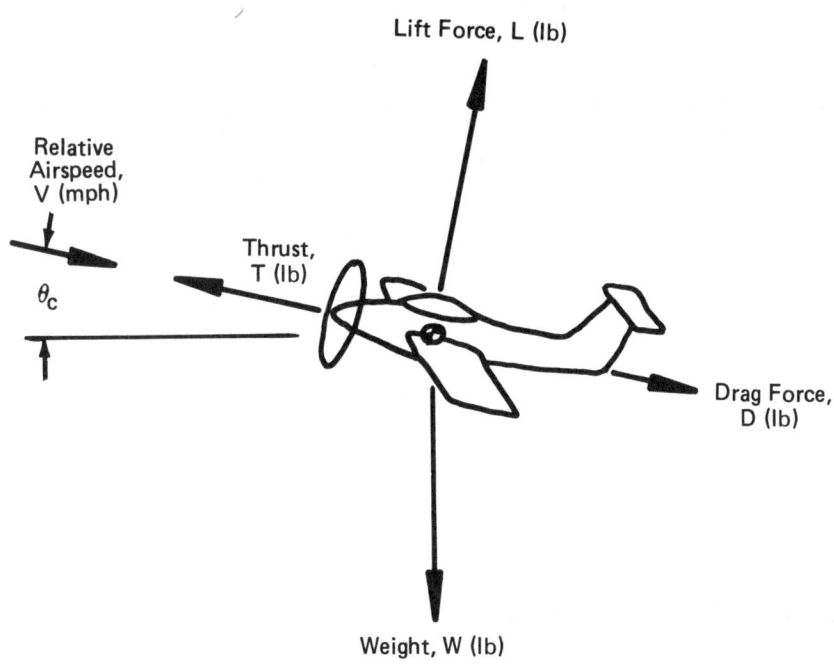

Figure 71. Force Balance for Climbing Flight in Equilibrium.

IDEALIZED PROPELLER THEORY

As a first estimate, let us calculate the propulsive efficiency of an idealized propeller using momentum theory. The propeller is replaced by disk which is assumed to increase the velocity of the air which passes through it. The change in momentum of this air mass is balanced by the force of the pressure acting at the propeller disk, which in turn provides the thrust to propel the airplane. In order to find the propulsive efficiency, we need to find the thrust power that is available from this idealized theory and divide by the power input.

The features of the analysis are shown in the schematic drawing in Figure 72. A mass of air enters the streamtube at the left with velocity V at a freestream pressure p_∞. As the air is accelerated through the propeller disk, the velocity increases to a value V_p and the streamtube area decreases to a value A_p at the propeller disk equal to the circular area swept by the propeller blade. The pressure decreases as the air is accelerated until it reaches a value p_1 just upstream of the propeller disk. It is then assumed that the pressure jumps to a value p_2 just downstream of the propeller, resulting in a thrust force equal to the pressure jump times the disk area, $T = (p_2 - p_1) A_p$. As the compressed air behind the propeller expands back to atmospheric pressure, p_∞, the velocity increases to a value V_3 and the streamtube area decreases to a value A_3 to maintain the same mass flux. The analysis neglects the effects of wind and viscosity and the precise, but complicated, aerodynamics in the neighborhood of the propeller blades. The objective is to obtain a simplified analysis that will give the scaling of propeller efficiency with altitude, engine size, propeller diameter and airspeed.

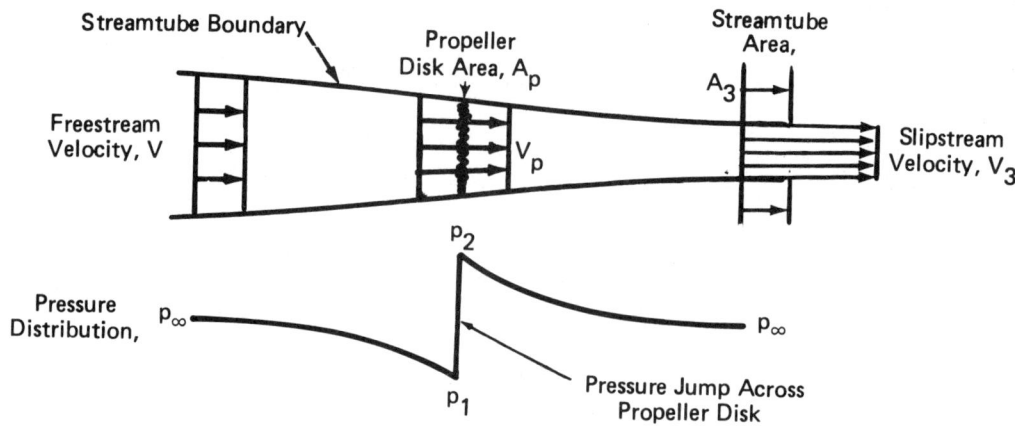

Figure 72. Schematic Diagram of an Idealized Propeller.

PROPELLER EFFICIENCY

The mass conservation equation — or continuity equation — states that the mass flow rate, $\dot{m}$, in the streamtube is constant. This is given by

$$\dot{m} = \rho A_p V_p = \rho A_3 V_3 \tag{39}$$

The momentum equation relates the change in momentum, $\dot{m}(V_3 - V)$, to the thrust force (equal to the pressure jump, $p_2 - p_1$, times the propeller disk area, A_p), $T = (p_2 - p_1) A_p$

$$\dot{m}(V_3 - V) = (p_2 - p_1) A_p = T \quad . \tag{40}$$

Bernoulli's equation relates the static and dynamic pressure to the total pressure in incompressible flow. Since we are increasing the energy of the air when it crosses the propeller disk, the total pressure will increase as the air crosses the plane of the propeller. Therefore, upstream of the propeller, we have

$$p_\infty + \frac{1}{2}\rho V^2 = p_1 + \frac{1}{2}\rho V_p^2 \quad . \tag{41}$$

Downstream of the propeller we have

$$p_2 + \frac{1}{2}\rho V_p^2 = p_\infty + \frac{1}{2}\rho V_3^2 \quad . \tag{42}$$

If we solve for $p_2 - p_1$ from (41) and (42) we have

$$p_2 - p_1 = \frac{1}{2}\rho(V_3^2 - V^2) \quad , \tag{43}$$

so that the thrust force is given by

$$T = \frac{1}{2}\rho(V_3 - V)(V_3 + V)A_p \qquad (44)$$

Comparing to (40), using (39) we find that

$$V_p = \frac{1}{2}(V_3 + V) \; , \qquad (45)$$

or

$$V_3 = 2V_p - V \; . \qquad (46)$$

Substituting (46) into (44) we find that the available propeller thrust is

$$T = 2\rho A_p V_p (V_p - V) \; . \qquad (47)$$

If it is assumed that no power is lost in the conversion of the torque at the shaft to the thrust at the propeller disk, then the shaft power is equal to the thrust times the velocity at the propeller disk, V_p. The usable power, however, is proportional to the thrust times freestream velocity, V. Therefore, the propulsive efficiency, η (which is equal to the thrust power divided by the shaft power), is proportional to the velocity ratio V/V_p

$$\eta = \frac{P_{thrust}}{P_{shaft}} = \frac{TV}{TV_p} = \frac{V}{V_p} \; . \qquad (48)$$

The engine power at the shaft is

$$P_{shaft} = 2\rho A_p V_p^2 (V_p - V) \qquad (49)$$

where consistent units have been used. If we are careful with units and express the airspeed in mph the engine power in brake horsepower, use the density ratio, $\sigma = \rho/\rho_{SL}$, together with the value of the sea level density, and express the propeller disk area in terms of the propeller diameter, we can rewrite (49) as

$$BHP = \frac{\pi}{2}\left[.002377\frac{(88)^3}{33000\,(60)^2}\right]\sigma D_p^2 \, V^3 \, \frac{1-\eta}{\eta^3} \; . \qquad (50)$$

Therefore, for a given brake horsepower, propeller diameter, and altitude (or density ratio), we have an expression for the propeller efficiency as a function of the velocity. If we define a characteristic propeller velocity, V_{prop}, by

$$V_{prop} = \underbrace{\left[\frac{33000\,(60)^2}{.002377\,(88)^3}\right]^{1/3}}_{41.86} \left[\frac{BHP}{\sigma D_p^2}\right]^{1/3} \quad (mph) \qquad (51)$$

(as given in relation ⑫), then the dimensionless velocity, $\tilde{V} = V/V_{prop}$, is related to the propeller efficiency by rearranging (50) to find

$$\tilde{V} = \left(\frac{2}{\pi}\right)^{1/3} \frac{\eta}{(1-\eta)^{1/3}} \quad . \tag{52}$$

This expression is plotted in Figure 38 and forms the basis for the template used to find the idealized rate of climb in equation (38). The practical way that this is used is discussed in Part 1.

If (52) is rearranged as a cubic equation for η, we have

$$\eta^3 + \left(\frac{\pi}{2}\tilde{V}^3\right)\eta - \left(\frac{\pi}{2}\tilde{V}^3\right) = 0 \quad . \tag{53}$$

This cubic equation can be solved using Appendix K, to give

$$\eta = \left(\frac{\pi}{4}\right)^{1/3} \tilde{V} \left\{ \left[1 + \sqrt{1 + \frac{2\pi}{27}\tilde{V}^3}\right]^{1/3} - \left[-1 + \sqrt{1 + \frac{2\pi}{27}\tilde{V}^3}\right]^{1/3} \right\} \tag{54}$$

When $\tilde{V} = 1$ (or when $V = V_{prop}$), the ideal propeller efficiency is η equal to 0.741.

In practice, the propeller efficiency will be smaller than this idealized value. This is especially true at the higher speeds where the propeller efficiency peaks and then decreases. This corresponds to speeds at which the local flow angle at the propeller blade decreases. Problems also appear in static conditions, wherein the propeller blades may stall (like a wing at too high an angle of attack).

The actuator disk theory presented in this section is an idealization for a constant speed propeller. In order to analyze the performance for a real propeller, whether fixed or variable pitch, many more parameters must be taken into account. These include the radial distribution of blade chord, pitch, airfoil section (thickness ratio, section lift/drag characteristics, etc.), and the effects of local Mach number, blade interference, tip loss factors, and fuselage interference.

For preliminary design purposes, the present theory is adequate. However, a more detailed performance analysis will require a closer look at propeller performance. (See the reference list for further reading).

ADVANCE RATIO, POWER COEFFICIENT AND NONDIMENSIONAL VELOCITY $\tilde{V}$

The nondimensional advance ratio, J, is defined by the forward speed of the propeller, V (ft/sec), the propeller rotational speed, n (revolutions/second), and the propeller diameter, D_p (ft):

$$J = \frac{V}{nD_p} \quad . \tag{55}$$

When the velocity is given in miles per hour and the rotational speed is given in RPM,

$$J = \frac{88 V}{(RPM) D_p} \quad . \tag{56}$$

J is a measure of the helix angle that the propeller blade makes as it rotates through the air. Together with the local propeller blade angle, the advance ratio will give the local angle of attack of the propeller blade element.

The dimensionless power coefficient, C_p, is related to the engine shaft power, P (ft-lb/sec); the free-stream density, ρ (slugs/ft^3 or lb sec^2/ft^4); propeller rotational speed, n (revolutions/sec); and propeller diameter, D_p (ft), by

$$C_p = \frac{P}{\rho n^3 D_p^5} \quad . \tag{57}$$

If the engine power is expressed in horsepowers BHP (hp), the density in terms of the density ratio, σ, and the sea level density (ρ_{SL} = 0.002377 slugs/ft^3) and rotational speed in RPM,

$$C_p = \frac{550 \, BHP \, (60)^3}{0.002377 \, (RPM)^3 \, D_p^5} \tag{58}$$

or

$$C_p = 5.00 \times 10^{10} \frac{BHP}{RPM^3 \, D_p^5} \tag{58a}$$

Therefore, we can eliminate the rotational speed by forming the ratio, $J/C_p^{1/3}$, and we find that

$$\frac{J}{C_p^{1/3}} = \tilde{V} = \frac{V}{V_{prop}} \quad . \tag{59}$$

Therefore, we see that the characteristic propeller speed, V_{prop}, is intimately related to the more traditional propeller parameters J and C_p. This also allows us to compare the simplified theory with actual propeller data by plotting η vs $J/C_p^{1/3}$. The optimized propeller data taken from Figure 3-20 (based on a Boeing Airplane Company General Propeller Chart) in Perkins and Hage (1949) is also plotted on Figure 38. This would correspond to an idealized variable pitch propeller that would constantly adjust itself so as to be optimally efficient. A fixed pitch propeller will have an efficiency that is lower at low and high forward speeds. At the design speed — whether a cruise prop or a climb prop — the efficiency will be as good as the variable pitch propeller, but at off-design conditions (a cruise prop at climb, or vice versa) the fixed pitch propeller will be less efficient.

This brief introduction to propeller theory will allow us to make first estimates of the power available for engine-propeller combinations. For more detailed aspects of the propeller selection process concerning blade area distribution and pitch angle, the reader is advised to consult Weick (1930) and advanced books on propeller design or Wood (1963) and other detailed design books.

STATIC THRUST

An approximation to the static thrust can also be made by returning to Equations (47) and (49). If we take V as equal to zero, solve for V_p from Equation (49) and substitute into Equation (49), we find that

$$T_s = \left(\frac{\pi}{2}\right)^{1/3} \rho^{1/3} D_p^{2/3} P_{shaft}^{2/3} . \qquad (60)$$

If we express the propeller diameter in feet and shaft power in HP, the idealized static thrust is

$$T_s = 10.41 \, \sigma^{1/3} (D_p \, BHP)^{2/3} \quad (lb) . \qquad (61)$$

We see that larger propellers with the same power input will generate larger static thrust. Equation (61) is given as part of relation ⑫ on the nomogram. The actual value of the static thrust will be less than this value, especially for a high pitched propeller, since propeller blades could be stalled at low forward speeds.

The ideal thrust available from an engine-propeller combination can be found from the propeller efficiency and the dimensionless speed $\tilde{V} = V/V_{prop}$,

$$\frac{T}{T_s} = \left(\frac{2}{\pi}\right)^{1/3} \frac{\eta}{\tilde{V}} = (1-\eta)^{1/3} \qquad (62)$$

If we substitute (54) into (62) we find

$$\hat{T} = \frac{1}{2^{1/3}} \left\{ \left[1 + \sqrt{1 + \frac{2\pi}{27}\tilde{V}^3} \right]^{1/3} - \left[-1 + \sqrt{1 + \frac{2\pi}{27}\tilde{V}^3} \right]^{1/3} \right\} \qquad (63)$$

where $\hat{T}$ is defined as T/T_s, the thrust divided by the ideal static thrust. The idealized thrust ratio is plotted as a function of velocity ratio in Figure 73, together with a line representing a value equal to 85% of the idealized thrust. This is the value that can probably be obtained by an optimally loaded propeller. Reference should be made to the advanced books mentioned earlier for methods of calculating the available thrust for a real propeller.

The curve for thrust available is mainly useful for takeoff performance analysis, where it is important to be able to calculate the acceleration as a function of speed. The acceleration is the thrust minus the drag (including rolling resistance of the wheels) all divided by the mass of the airplane.

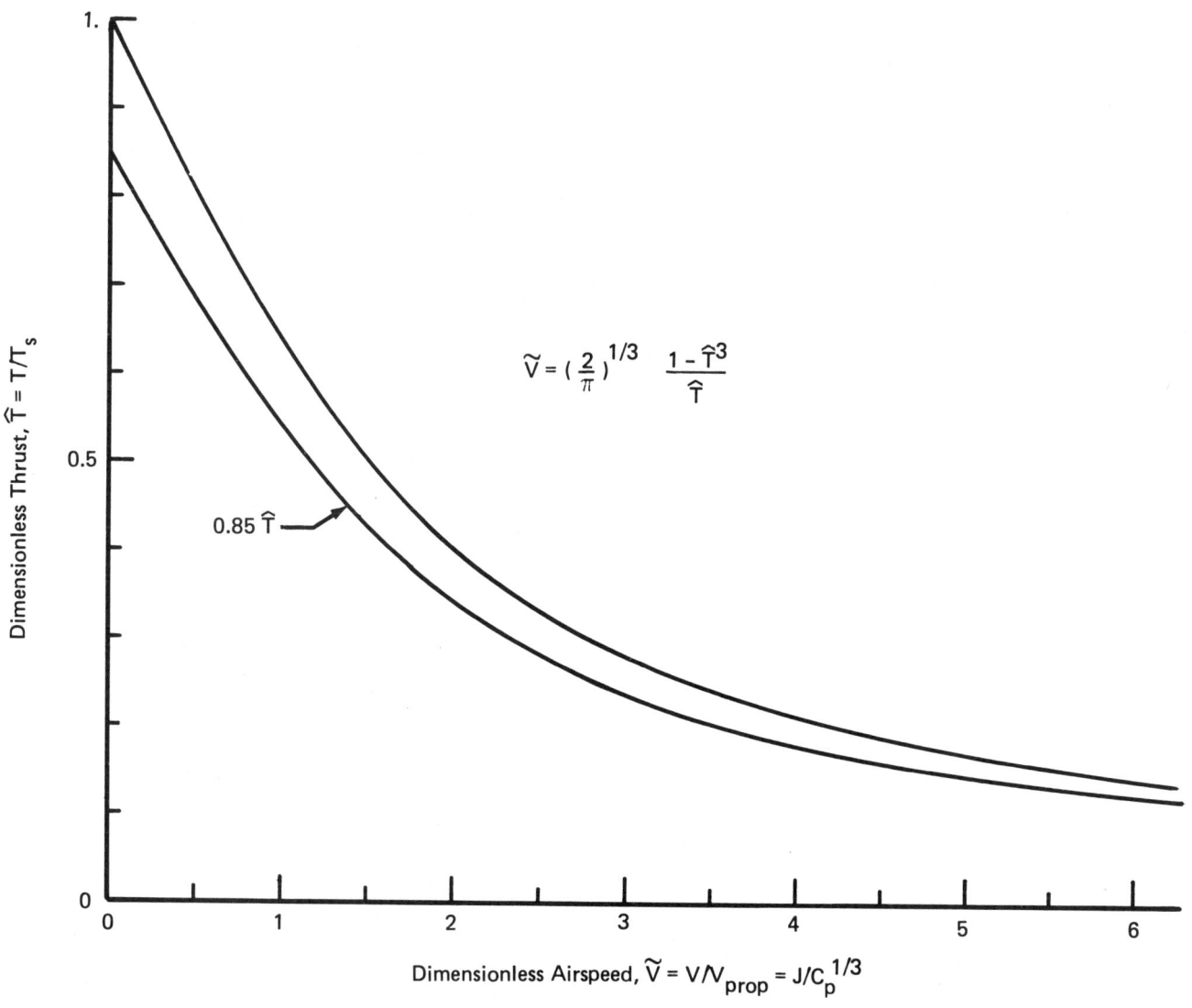

Figure 73. Idealized Thrust as a Function of Airspeed.

PROPELLER TIP SPEED

If the propelller diameter is made larger, we expect to obtain better efficiency and more static thrust. However, if the tip speed approaches the speed of sound, the losses associated with compressibility effects reduce the propeller efficiency considerably. It is therefore desirable to keep the propeller tip Mach number — $M_p = V_{tip}/a$, where V_{tip} is the propeller tip speed in ft/sec and a is the speed of sound — lower than 0.8. This corresponds to a tip speed of 880 ft/sec if the speed of sound is taken to be 1100 ft/sec. At tip speeds above this, the efficiency drops off rapidly and the noise level rises dramatically. The propeller rotational speed, RPM, propeller diameter, D_p (ft) and tip Mach number are related by

$$M_p = \frac{\pi D_p \text{ RPM}}{(60)(1100)} = \frac{D_p \text{ RPM}}{21,008} \tag{64}$$

given in relation ⑬ on the nomogram.

This relation is to be used as a check so that the propeller diameter and rotational speed result in a small enough propeller tip Mach number to minimize tip losses and noise.

SUMMARY OF PERFORMANCE RELATIONS

Two conditions are of interest to designers of airplanes and sailplanes. The first is the condition which yields the minimum sink rate for sailplanes and the minimum power required for level flight for an aircraft. This occurs at a relative airspeed, V_{minS}. The second is the condition for maximum lift-to-drag ratio for sailplanes and minimum drag for powered aircraft (corresponding to speed for maximum range if the power available is constant). This occurs at a relative speed which is $3^{1/4}$ or 1.32 times higher than V_{minS}. The expressions for airspeed, lift coefficient, power required, sink rate, drag and lift-to-drag ratio are tabulated in Table 2 for the appropriate condition in terms of the weight, effective span, drag area, and effective chord. The characteristic propeller velocity and the idealized static thrust are also tabulated in terms of the propeller diameter and available shaft power. The formulas are gathered here for easy reference and represent the information contained in the Airplane Performance and Design Nomogram.

Table 2. Summary of Airplane Performance Relations.

① $W/S = \dfrac{\sigma C_L V^2}{391}$ (lb/ft²)   [V(mph)]

⑤ $THP_a = \eta\, BHP \approx \dfrac{\sigma A_D V_{max}^3}{146625}$ (hp)   [A_D(ft²)]

⑦ $V_{minS} = 11.29 \dfrac{\sqrt{W/b_e}}{\sqrt{\sigma}\, A_D^{1/4}}$ (mph); $\widehat{V} = V/V_{minS}$

⑦ $THP_{min} = 0.03921\, A_D^{1/4} \dfrac{(W/b_e)^{3/2}}{\sqrt{\sigma}}$ (hp)   [W(lb); b_e(ft)]

⑦ $D_{min} = 1.128\, \sqrt{A_D}\, \dfrac{W}{b_e}$ (lb)

⑧ $R_{S,min} = 33{,}000\, \dfrac{THP_{min}}{W} = 1294.0\, A_D^{1/4} \dfrac{\sqrt{W}}{\sqrt{\sigma}\, b_e^{3/2}}$ (fpm); $\widehat{R}_S = R_S/R_{S,min}$

⑨ $(L/D)_{max} = 0.8862 \dfrac{b_e}{\sqrt{A_D}}$; $V_{maxLD} = 1.316\, V_{minS}$

⑩ $C_{L,minS} = 3.07 \dfrac{\sqrt{A_D}}{c_e}$

⑪ $R^*_{C,max} = 33{,}000\, \dfrac{BHP}{W}$ (fpm); $\eta = THP_a/BHP$

⑫ $V_{prop} = 41.8 \left[\dfrac{BHP}{\sigma D_p^2}\right]^{1/3}$ (mph)   [D_p(ft)]; $\widetilde{V} = V/V_{prop}$

⑫ $T_s = 10.41\, [\sigma D_p^2\, BHP^2]^{1/3}$ (lb); $\widehat{T} = T/T_s$

⑬ $M_p = \dfrac{RPM\, D_p}{21{,}008}$

$$\widehat{R}_S = \dfrac{\widehat{V}^3}{4} + \dfrac{3}{4\widehat{V}} \qquad \widetilde{V} = \eta \left[\dfrac{2/\pi}{1-\eta}\right]^{1/3} \qquad \widehat{T} = \left(\dfrac{2}{\pi}\right)^{1/3} \dfrac{\eta}{\widetilde{V}}$$

$$\eta = \left(\dfrac{\pi}{4}\right)^{1/3} \widetilde{V} \left\{\left[1 + \sqrt{1 + \dfrac{2\pi}{27}\widetilde{V}^3}\right]^{1/3} - \left[-1 + \sqrt{1 + \dfrac{2\pi}{27}\widetilde{V}^3}\right]^{1/3}\right\}$$

APPENDIX

A. Abbreviations and Symbols

B. What is a Nomogram?

C. Discussion of Units

D. Standard Atmosphere

E. FORTRAN Computer Program for Performance Analysis

F. Airplane Efficiency Factor, e; Ground Effect

G. Drag Analysis

H. Airfoil Selection

I. Reynolds Number

J. Equation of State

K. Solution of Cubic Equation

L. Tabulated Performance Data for Various Aircraft

M. How to Calculate Drag Area, A_D, and Airplane Efficiency Factor, e, from Flight Test Data   p. 182

APPENDIX A. ABBREVIATIONS AND SYMBOLS

The following abbreviations are used for the units of length, area, volume, time, force, mass, speed, power, pressure, temperature and density:

ft	Length in feet
mi	Length in statute miles (5280 ft)
ft^2	Area in square feet
ft^3	Volume in cubic feet
sec	Time in seconds
min	Time in minutes
hr	Time in hours
lb	Force in pounds
lb_m	Pound-mass (mass which weighs 1 lb in the Earth's gravitational field)
slug	Fundamental unit of mass in English Engineering System of units (32.2 lb_m; lb sec^2/ft)
fpm	Speed in feet per minute (ft/min)
mph	Speed in miles per hour (mi/hr)
radian	Angular measure (2π radians = 360 degrees)
RPM	Angular speed in revolutions per minute
hp	Power in horsepower (550 ft lb/sec; 33000 ft lb/min)
lb/ft^2	Pressure in pounds per square foot
F	Temperature in degrees Fahrenheit
R	Temperature in degrees Rankine (= degrees F + 460)
$slug/ft^3$	Density in slugs per cubic feet

The symbols used in the text and figures are listed in alphabetical order. The units of the defined quantity are given in parentheses in modified English Engineering units.

a	Local sound speed (ft/sec)
A_D	$= C_{D,O} S$, drag area (ft^2)
$A_{D,\pi}$	Drag area of individual aircraft components (ft^2) [Appendix G]
A_f	Projected frontal area (ft^2)
A_p	Propeller disk area (ft^2) [Figure 72]
AR	$= b^2/S = b/c$, aspect ratio
b	Wing span (ft)
b_e	$= b\sqrt{e}$, effective span (ft)
BHP	Brake horsepower (hp)
c	$= S/b$ average chord of the wing (ft)
C_D	Drag coefficient (–) [Equation (4)]
$C_{D,f}$	Drag coefficient based on frontal area (–) [Appendix G]
$C_{D,i}$	Induced drag coefficient (–) [Equation (13)]
$C_{D,w}$	Drag coefficient based on wetted area (–) [Appendix G]
$C_{D,\pi}$	Drag coefficient based on the characteristic area S_π for a particular aircraft component (–) [Appendix G]
$C_{D,O}$	Zero-lift drag coefficient (–)
c_e	$= c/\sqrt{e}$, effective chord (ft)
C_L	Lift coefficient (–) [Equation (3)]
$C_{L,max}$	Maximum lift coefficient (–)
$C_{L,maxLD}$	Lift coefficient at maximum lift-to-drag ratio [Equation (27)]
$C_{L,minS}$	Lift coefficient at minimum sink conditions (–) [Equations (18) and (19)]
C_p	Power coefficient for engine-propeller combination (–) [Equation (58)]

D	Drag force (lb)
D_{min}	Minimum drag (lb) [Equation (30)]
D_p	Propeller diameter (ft)
e	Airplane efficiency factor (–) [Appendix F]
e_w	Wing efficiency factor (–) [Figure F.1]
eAR	Effective aspect ratio (–)
F_p	$= \dfrac{R_{C,max} W_u}{33000\ BHP}\left[1 - \dfrac{V_{min}}{V_{max}}\right]$, performance rating parameter (–)
h	Density Altitude (ft) [Appendix D]
J	Advance ratio, 88 V/RPM D_p (–) [Equation (56)]
L	Lift force (lb)
L/D	Lift-to-drag ratio (–)
$(L/D)_{max}$	Maximum lift-to-drag ratio [Equations (28) and (29)]
$\dot{m}$	Mass flow rate through the propeller disk (slugs/sec) [Equation (39)]
M_p	$= V_{tip}/a$, propeller tip Mach number (–) [Equation (64)]
R_C	Rate of climb (fpm) [Equation (38)]
$R_{C,max}$	Maximum rate of climb (fpm)
R_C^*	$= \eta R_{C,max}^*$, Ideal rate of climb (fpm)
$R_{C,max}^*$	$= 33000\ BHP/W$ Maximum ideal rate of climb (fpm)
R_S	$= V \sin \theta_g$, sink rate (fpm)
$R_{S,min}$	Minimum sink rate (fpm) [Equation (20)]
$\hat{R}_S$	Nondimensional sink rate (–) [Equation (25)] *
p_1	Pressure just upstream of propeller disk (lb/ft^2) [Figure 72]
p_2	Pressure just downstream of propeller disk (lb/ft^2) [Figure 72]

*The circumflex (⌢) and tilde (~) represent nondimensional quantities.

p_∞	Freestream pressure (lb/ft^2) [Figure 72]
RPM	Propeller shaft speed (revolutions/min)
S	Wing area, (ft)
S_w	Wetted area (ft^2)
S_π	Characteristic reference area for drag of a particular aircraft component (ft^2) [Appendix G]
SFC	Specific fuel consumption (gal/hp hr)
T	Thrust (lb) [Figure 70]
$\hat{T}$	= T/T_s, dimensionless thrust [Equation (63) and Figure 73] *
T_s	Static thrust (lb) [Equation (61)]
THP$_a$	Thrust horsepower available (hp)
THP$_{a,L}$	Thrust horsepower available for level flight (hp) [Equation (31)]
THP$_{min}$	Minimum thrust horsepower required for level flight (hp) [Equation (33)]
V	Relative airspeed (mph)
$\hat{V}$	= V/V_{minS}, nondimensional velocity (-) [Equation (25)] *
$\tilde{V}$	= V/V_{prop} = $J/C_p^{1/3}$, nondimensional velocity (-)*
V_c	Cruise velocity (mph)
V_{max}	Maximum level flight speed (mph)
V_{maxLD}	= 1.316 V_{minS}, airspeed for maximum L/D ratio (mph)
V_{min}	Minimum level flight speed (mph)
V_{minS}	Airspeed for minimum sink rate (mph) [Equation (21)]
V_p	Slipstream velocity at the propeller disk (ft/sec) [Figure 72]
V_{prop}	Characteristic propeller velocity (mph) [Equation (51)]
$V_{s,O}$	Stall speed in landing configuration (mph)

$V_{s,1}$	Stall speed in clean configuration (mph)
V_{tip}	Speed of the tip of the propeller (ft/sec)
V_x	Speed for the best angle of climb (mph)
V_y	Speed for the best rate of climb (mph)
V_3	Slipstream velocity far downstream of the propeller disk (ft/sec) [Figure 72]
W	Gross weight (lb)
W_e	Empty weight (lb)
W_u	$= W - W_e$, useful load (lb)
W/b_e	Effective span loading (lb/ft)
W/S	Wing loading (lb/ft^2)

Greek Symbols

α (alpha)	Angle of attack (–) [Appendix H]
β (beta)	Adiabatic lapse rate (R/ft) [Appendix D]
η (eta)	$= THP_a/BHP$, Propeller efficiency (–)
θ_c (theta)	Climb angle (degrees) [Figure 71]
θ_g	Glide angle (degrees) [Figure 68]
θ_{max}	Maximum climb angle (degrees)
π (pi)	= 3.1415926 (ratio of the circumference of a circle to its diameter)
ρ (rho)	Local freestream density (slugs/ft^3)
ρ_{SL}	Density at standard sea level conditions (0.002377 slugs/ft^3)
σ (sigma)	$= \rho/\rho_{SL}$, density ratio (–)

APPENDIX B. WHAT IS A NOMOGRAM?

A nomogram, according to Webster's New Collegiate Dictionary, is "... a graphic representation that consists of several lines marked off to a scale and arranged in such a way that by using a straightedge to connect known values on two lines an unknown value can be read at the point of intersection with another line." The root word is "nomos" a Greek word meaning law. Nomograms are of many types, but those used here are, strictly speaking, called alignment charts. Construction techniques and the mathematical theory for nomograms are discussed more completely in books such as that by Douglass and Adams (1947). In this appendix we will discuss how a nomogram works, in general.

The usefulness of a nomogram is best illustrated by examples (Figure B.1) which show how to work problems in addition and multiplication. Figure B.1a can be used for the addition: A plus B equals C, where the scales of A, B, and C are linear and where the range of A and B is from zero to ten. The sample problem shows the sum of nine plus four (line ①) or four plus nine (line ②) equal to thirteen. If we change the range of B, so that the value of B lies between zero and twenty, the **location** of the C scale is changed, as shown in Figure B.1b. In addition, the **range** of the C scale changes.

The multiplication problem: a times b equals c, is considered in Figures B.1c and B.1d. If we first take logarithms of both sides of the equation, c = a.b, then the problem becomes one of addition (of logarithms). Thus, changing to a logarithmic scale, we can perform the multiplication directly. A change in the range of b has an effect similar to that discussed in the addition problem, that is, the **location** and **range** of the c-scale changes. The sample calculation illustrates the multiplication of three times six. The answer must be interpolated on the c-scale and is the same whether line ③ or line ④ is constructed.

The nomogram can easily handle all power-law relationships such as

$$c = K a^\alpha b^\beta , \quad (B.1)$$

where K, α, and β are constants. Taking logarithms, we obtain

$$\log c = \log K + \alpha \log a + \beta \log b . \quad (B.2)$$

The K factor has the effect of sliding the origin of the c-scale up or down its axis and the α and β terms affect the grid spacing on the a– and b–scales. Relation ① is an example of a power-law expression taken directly from the Airplane Performance and Design Nomogram. The wing loading, W/S, lift-coefficient, C_L, and sea level velocity, V, are related and by

$$W/S = \frac{C_L V^2}{391} \quad (\text{lb/ft}^2) \quad (B.3)$$

where the relative airspeed, V, is given in mph. Taking logarithms of both sides of (B.3), we have

$$\log W/S = \log C_L + 2 \log V - \log 391 . \quad (B.4)$$

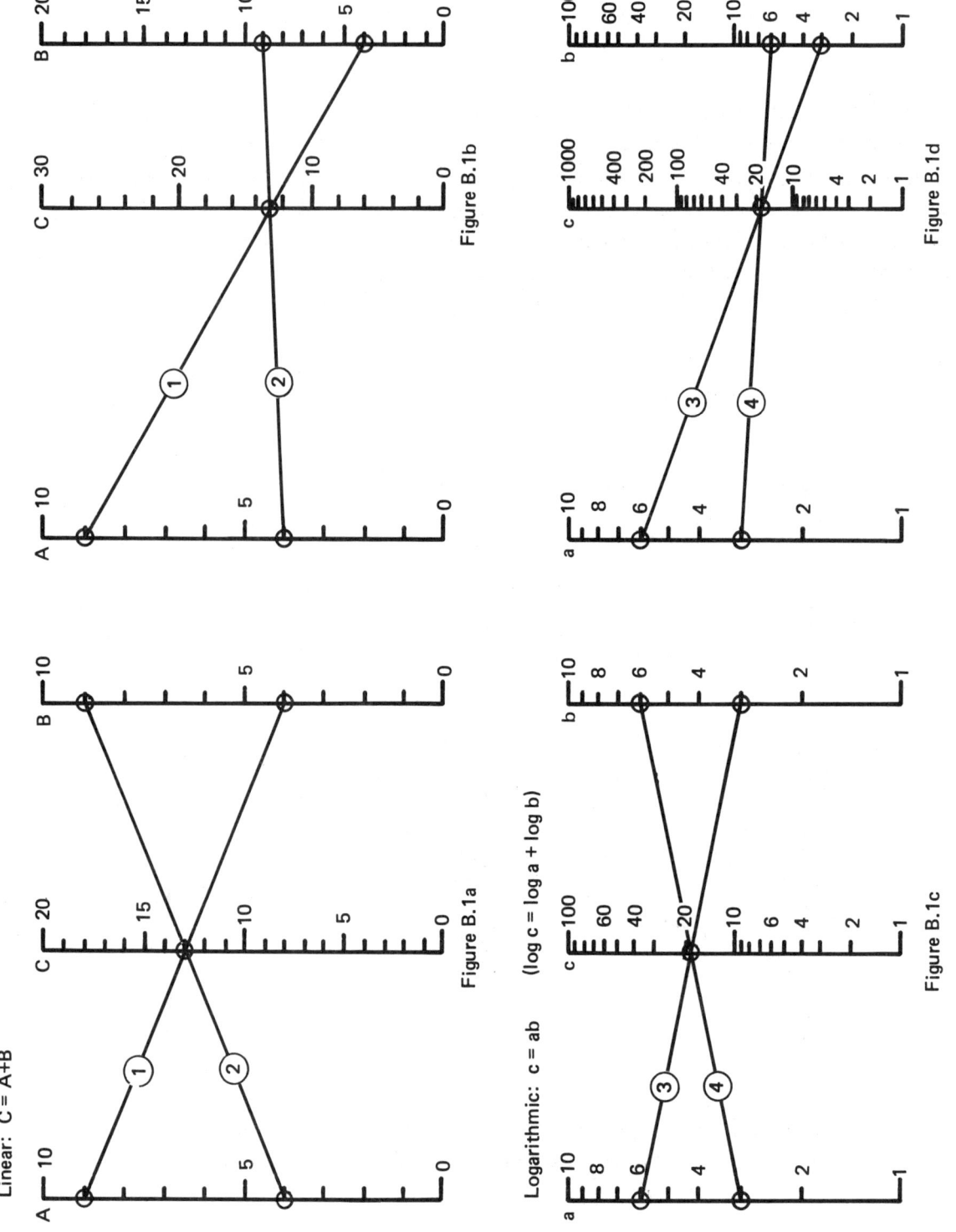

Figure B.1. Illustrative Example of a Nomogram.

If we rearrange (B.4), we have

$$\log V = \frac{1}{2} \log W/S - \frac{1}{2} \log C_L + \frac{1}{2} \log 391 \ . \tag{B.5}$$

Notice that we have rearranged the relationship so that C_L and W/S are now taken to be the independent variables. Since we are **subtracting** the logarithm of C_L in (B.5), notice that the scale for C_L on the nomogram in Figure B.2 is inverted. With this nomogram we can enter with C_L and W/S to obtain V. However, we can also enter with C_L and V to obtain W/S, or with W/S and V for C_L. This is one major advantage of the nomogram — arbitrary choice of independent variables. In this case we chose a maximum lift coefficient of 1.5 and a wing loading of 6.8 lb/ft² to find a stall speed of 42 mph.

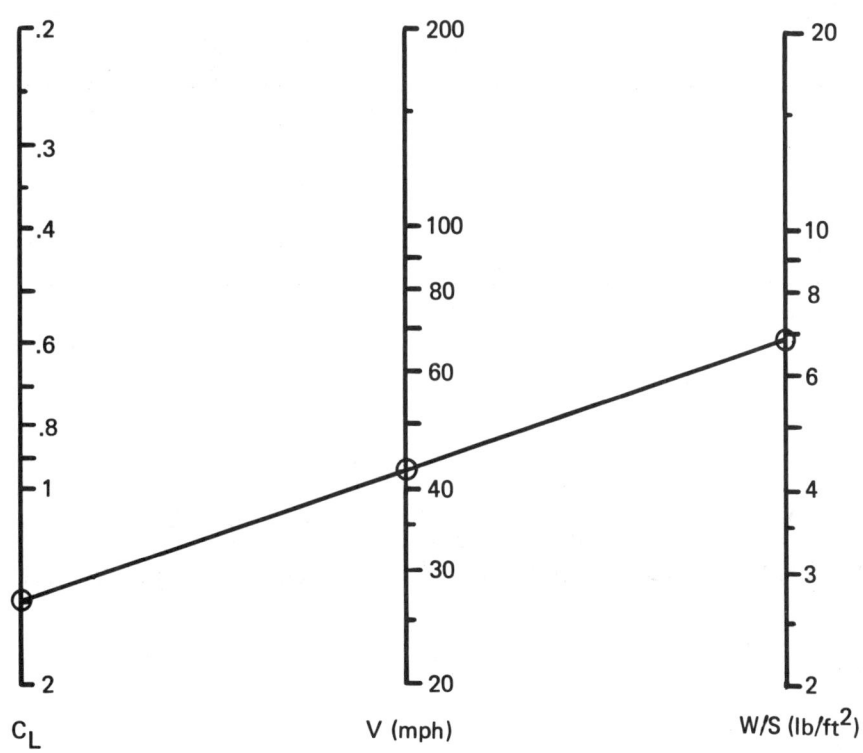

Figure B.2. Nomogram Relating Lift Coefficient, C_L, Airspeed, V, and Wing Loading, W/S, at Sea Level Conditions. (Relation ①).

APPENDIX C. DISCUSSION OF UNITS

When we are dealing with real-world engineering problems, it is important to be able to convert the results obtained with one system of units into answers in another set of units. This is done by using equivalent quantities. For instance, we know that 5280 feet is equal to one mile. Therefore, the quantity (5280 ft/mi) is equal to unity. Similarly, the quantity (60 min/hr) is a quantity that is equal to unity. Since we can multiply or divide any quantity by the number 1.0 and not change the value, we can multiply or divide by the bracketed numbers (with their dimensional characteristics) and not change the physical meaning of the quantity of interest. We can illustrate this by converting a speed, say 150 miles per hour, into the same speed given in feet per minute. First, write the speed, V, as

$$V = \frac{150 \text{ mi}}{\text{hr}}$$

Since miles occurs in the numerator, multiply by the first conversion factor (5280 ft/mi). Then, divide by the second conversion factor (60 min/hr). By cancelling the dimensions of miles from the numerator and dominator, and doing the same for the dimensions of hours, we end up with a set of numbers to be computed that have the dimensions of feet per minute (fpm).

$$V = \frac{150 \text{ mi}}{\text{hr}} \left[\frac{5280 \text{ ft}}{\text{mi}} \right] \left[\frac{\text{hr}}{60 \text{ min}} \right] = \frac{(150)(5280)}{60} \frac{\text{ft}}{\text{min}} = 13,200 \text{ fpm} .$$

The terms in the square brackets are equal to one so the value of the expression has not changed. That is, 13,200 fpm is equal to 150 mph. We have listed in Table C-1 several sets of conversion factors that may be useful in the calculation of the performance-related parameters for aircraft. The abbreviations used are defined in Appendix A.

The major difficulty with various systems of units seems to come from the confusion between mass and weight. When a body with a certain mass M is weighted in the Earth's gravitational field, where the acceleration of gravity is g, then the weight is equal to W according to Newton's second law

$$W = Mg$$

In this way, a mass of 1 lb_m is defined as a weight of 1 lb under the influence of gravity, g = 32.2 ft/sec^2. That means that

$$1 \text{ lb} = (1 \text{ lb}_m)(32.2 \text{ ft/sec}^2)$$

or, in a rearranged form

$$\left[\frac{32.2 \text{ lb}_m \text{ ft}}{\text{lb sec}^2}\right] = 1 \ .$$

The factor in brackets is a factor that is equal to unity so that it can be multiplied by or divided into any quantity without changing its physical value. Remember that lb is a pound-force and lb_m is a pound-mass. Difficulties with conversion factors can be avoided by working in a "consistent" set of units. One such set of units is the English engineering set of units where the unit of mass is the "slug" (not too flattering a term). When the slug is used for the unit of mass, Newton's second law

$$F = Ma$$

reduces to a simple form: a force, F, of 1 pound will accelerate a mass, M, of 1 slug at a rate, a, of 1 foot per second per second, or

$$1 \text{ lb} = (1 \text{ slug}) (1 \text{ ft/sec}^2) \ .$$

Comparing this expression with that above, we find that the mass of 1 slug is equivalent to a mass that would weigh 32.2 lb at the Earth's surface, or

$$1 \text{ slug} = 32.2 \text{ lb}_m \ .$$

Definitions

$$\left[\frac{\text{slug ft}}{\text{lb sec}^2}\right] = 1 \ , \quad \left[\frac{32.2 \text{ lb}_m \text{ ft}}{\text{lb sec}^2}\right] = 1 \ , \quad \left[\frac{\text{N sec}^2}{\text{kg m}}\right] = 1 \ ,$$

$$\left[\frac{33000 \text{ ft lb}}{\text{hp min}}\right] = 1 \ , \quad \left[\frac{550 \text{ ft lb}}{\text{hp sec}}\right] = 1 \ , \quad \left[\frac{\text{W sec}}{\text{N m}}\right] = 1$$

Conversion Factors

1 ft = 0.3048 m   1 lb = 4.448 N

1 slug = 14.594 kg   1 hp = 746 W

1 lb_m = 0.4536 kg   60 mph = 88 $\frac{\text{ft}}{\text{sec}}$

---

m: meter,   N: Newton,   W: Watt,   kg: kilogram

Table C.1. Units: Definitions and Conversion Factors

APPENDIX D. STANDARD ATMOSPHERE

In the calculation of airplane performance it is important to know the variation of the density as a function of altitude. The density appears in the definitions of the lift and drag coefficients, governs the brake power-altitude factor, $\phi(h)$, and is related to the propeller efficiency through the quantity V_{prop}, the airspeed for a 74% ideal propeller efficiency.

It is not the actual geometric altitude that is of primary importance, but the density altitude or, as Richard Taylor (1977) calls it, the performance altitude. In this section, we will derive an approximate expression for the variation of the density ratio, σ, with respect to the density altitude, h.

We first take the differential form for the vertical momentum equation where we relate the relative change of pressure with respect to altitude, dp/dh, to the local density, ρ, and gravitational acceleration, g:

$$\frac{dp}{dh} = -\rho g \, , \tag{D.1}$$

where the coordinate h is pointed up. If the density were constant and equal to ρ_0, we could integrate this equation directly to get

$$p = p_0 - \rho_0 g h \, , \tag{D.2}$$

which is the hydrostatic variation that we would find in a swimming pool, say (where p increases when h decreases and ρ_0 would be the density of water).

For a gas such as air, the equation of state (Appendix J) is given by

$$p = \rho R T \, , \tag{D.3}$$

where R is a constant (1718 ft^2/sec^2 R) and the absolute temperature T is given in degrees Rankine (equal to the temperature in degrees Fahrenheit plus 460).

If this equation is substituted into the differential equation for the pressure, (D.1), we obtain

$$\frac{1}{p} \frac{dp}{dh} = -\frac{g}{RT_0} \tag{D.4}$$

For an isothermal atmosphere, the temperature remains constant, equal to T_0, and the equation can be integrated to give

$$\log \frac{p}{p_0} = -\frac{gh}{RT_0} \, , \tag{D.5}$$

or

$$p = p_O \exp(-gh/RT_O) . \tag{D.6}$$

The density ratio that would result if this expression were substituted back into (D.3) is

$$\sigma = \exp(-gh/RT_O) . \tag{D.7}$$

If we take T_O to be $T_{SL} = 519°R$ and if we define the characteristic altitude $h_O = RT_O/g = 27{,}700$ ft, then

$$\sigma = \exp(-h/h_O) \tag{D.8}$$

In the standard atmosphere, the temperature decreases with altitude at the standard lapse rate of 3.56 degrees Fahrenheit (or Rankine) per 1000 feet. Therefore, (D.4) can be written as

$$\frac{1}{p}\frac{dp}{dh} = \frac{-g}{R(T_O - \beta h)} \tag{D.9}$$

where $\beta = .00356$ (R/ft) is the lapse rate and $T_O = T_{SL} = 519$ R. If we integrate this equation, we obtain

$$\log \frac{p}{p_{SL}} = \frac{g}{R\beta} \log \left[1 - \frac{\beta h}{T_{SL}}\right] \tag{D.10}$$

or

$$\frac{p}{p_{SL}} = \left[1 - \frac{\beta h}{T_{SL}}\right]^{(g/R\beta)} \tag{D.11}$$

When this equation is substituted into (D.3), we have the equation for the variation of the density ratio with altitude

$$\sigma = \left[1 - \frac{\beta h}{T_{SL}}\right]^{(\frac{g}{R\beta} - 1)} \tag{D.12}$$

If the values for β, R, g, and T_{SL} are substituted into this equation, we have

$$\sigma = \left[1 - \frac{h}{145{,}800}\right]^{4.265} \tag{D.13}$$

where h is the altitude in feet. This is the expression plotted in Figure D.1, along with the approximation (D.8). Equation (D.13) holds up to an altitude of 36,240 feet, where the temperature is taken to be a constant at an average value of 390 R (-70 F). The equation for the density ratio in this region (36,240 ft to 82,000 ft) is

$$\sigma = 1.688 \exp\left[\frac{-h}{20{,}808}\right] \tag{D.14}$$

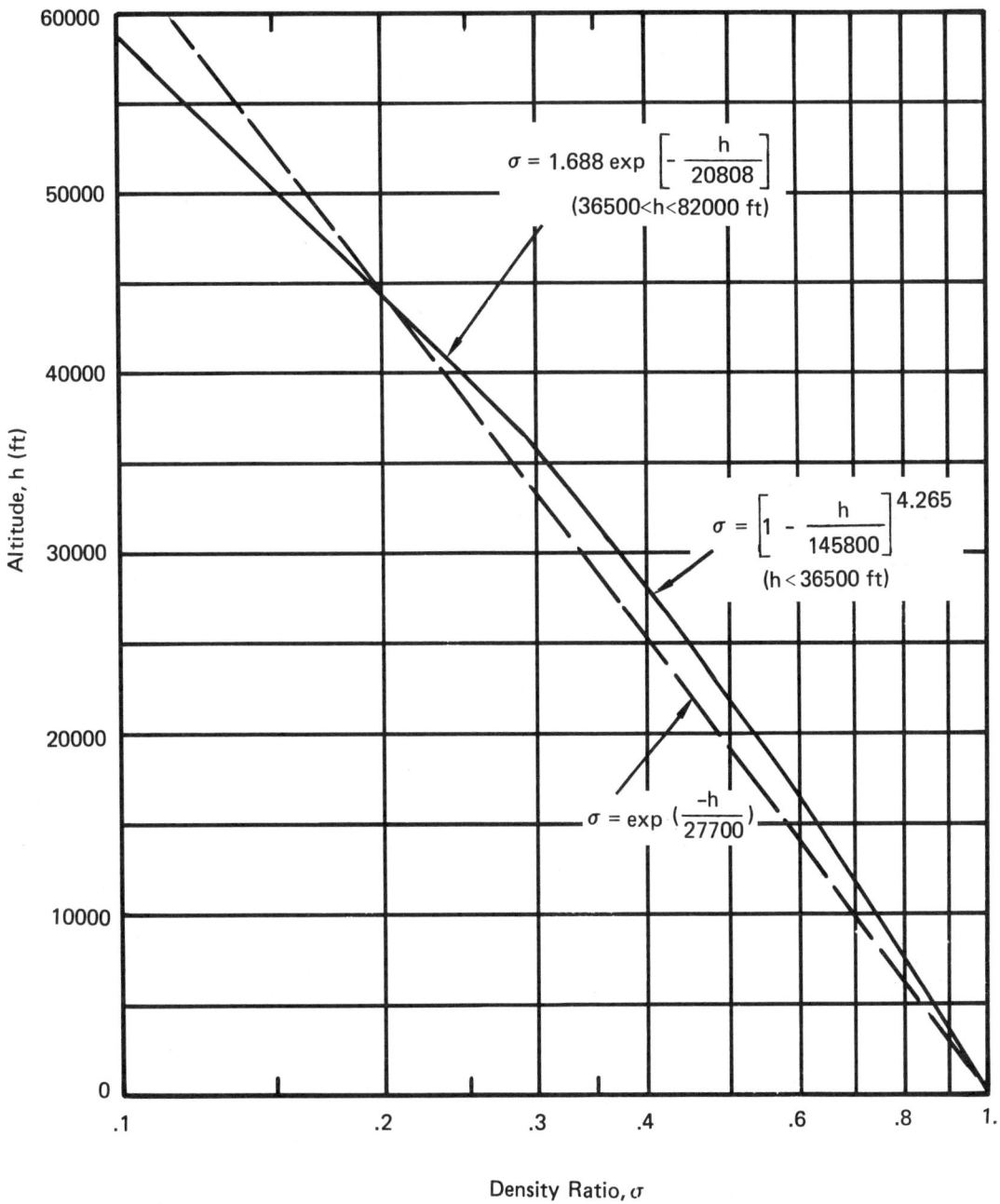

Figure D.1. Standard Atmosphere.

The properties of the standard atmosphere are given below

Sea-level pressure

$$p_{SL} = 2116.2 \, \frac{\text{lb}}{\text{ft}^2} = 14.69 \, \frac{\text{lb}}{\text{in}^2} = 1.013 \times 10^5 \, \frac{\text{N}}{\text{m}^2}$$

Sea-level temperature

$$T_{SL} = 58.7°F = 15°C$$

Sea-level density

$$\rho_{SL} = 0.002377 \, \frac{\text{slug}}{\text{ft}^3} = 0.002377 \, \frac{\text{lb sec}^2}{\text{ft}^4} = 1.224 \, \frac{\text{kg}}{\text{m}^3}$$

Table D.1. Standard Sea Level Atmospheric Conditions

APPENDIX E. FORTRAN COMPUTER PROGRAM FOR PERFORMANCE ANALYSIS

In this appendix we present a sample computer program for the airplane performance based on the equations derived in Part 2 of this book. The standard atmosphere as well as the calculation of the Reynolds number are also included. The airplane efficiency, e, and the drag area have to be determined from analyses similar to those of Appendix F and Appendix G. The listing is self-explanatory when studied with the program input and output for the sample calculation of the performance for the T-18.

Although the program is given in FORTRAN, the program can be easily converted to BASIC or other program languages. If there are questions about the programming, consult the equations themselves. The nomogram will give results that are as reliable as the computer program, however, since the uncertainties in the calculation of the drag area, airplane efficiency factor, and maximum lift coefficient more than offset the numerical inaccuracy of the nomogram.

```
C      PROGRAM TO CALCULATE AIRPLANE PERFORMANCE PARAMETERS.
C
C      INPUT PARAMETERS:
C
C      VS1       STALL SPEED WITHOUT FLAPS (MPH)
C      CLMAX     MAXIMUM LIFT COEFFICIENT
C      CLMAXF    MAXIMUM LIFT COEFFICIENT WITH FLAPS
C      W         GROSS WEIGHT (LB)
C      WU        USEFUL LOAD (LB)
C      B         WING SPAN (FT)
C      E         AIRPLANE EFFICIENCY FACTOR
C      BHP       BRAKE HORSEPOWER OF ENGINE (HP)
C      VMAX      DESIRED MAXIMUM LEVEL FLIGHT SPEED (MPH)
C      DP        PROPELLER DIAMETER (IN)
C      RPM       PROPELLER SPEED (RPM)
C      ALT       ALTITUDE (FT)
C      DELV      AIRSPEED INCREMENT FOR PERFORMANCE DATA (MPH)
C
       REAL LDMAX,MP
C
C      VALUES FOR INPUT PARAMETERS (THORP T-18 SAMPLE CALCULATION)
C
       VS1=67.
       CLMAX=1.53
       CLMAXF=2.1
       W=1500.
       WU=600.
       B=20.833
       E=0.744
       BHP=150.
       VMAX=180.
       DP=72.
       RPM=2700.
       ALT=0.
       DELV=10.
C
C      BEGIN CALCULATION
C
       WS=CLMAX*VS1*VS1/391.
       VS0=SQRT(WS*391./CLMAXF)
       S=W/WS
       AR=B*B/S
       C=B/AR
       BE=B*SQRT(E)
       EAR=BE*BE/S
       CE=S/BE
       WBE=W/BE
       AD=.8*BHP*146625./(VMAX**3)
       CDO=AD/S
       AD2=SQRT(AD)
       AD4=SQRT(AD2)
       VMINS=11.29*SQRT(WBE)/AD4
       THPM=0.03922*AD4*WBE*SQRT(WBE)
       RSMIN=33000.*THPM/W
```

```
      LDMAX=0.8862*BE/AD2
      DMIN=W/LDMAX
      CLMINS=3.07*AD2/CE
      RCSTAR=33000.*BHP/W
      DPF=DP/12.
      VTIP=RPM*DPF*.05236
      MP=VTIP/1100.
      VPROP=41.9*(BHP/DPF/DPF)**(1./3.)
      TS=10.41*(BHP*DPF)**(2./3.)
      PRINT 100,VS1,CLMAX,CLMAXF,W,WU,B,E,BHP,VMAX,DP,RPM,ALT
100 FORMAT(/19H   INPUT PARAMETERS://
    F 27H    STALL SPEED WITHOUT FLAPS,18X,6HVS1 = ,F6.1,4H MPH/
    F 26H    MAXIMUM LIFT COEFFICIENT,17X,8HCLMAX = ,F6.3/
    F 31H    MAXIMUM LIFT COEFF WITH FLAPS,11X,9HCLMAXF = ,F6.3/
    F 14H    GROSS WEIGHT,33X,4HW = ,F6.0,3H LB/
    F 13H    USEFUL LOAD,33X,5HWU = ,F6.0,3H LB/
    F 11H    WING SPAN,36X,4HB = ,F6.2,3H FT/
    F 28H    AIRPLANE EFFICIENCY FACTOR,19X,4HE = ,F6.3/
    F 25H    ENGINE BRAKE HORSEPOWER,20X,6HBHP = ,F6.0,3H HP/
    F 21H    MAXIMUM LEVEL SPEED,23X,7HVMAX = ,F6.1,4H MPH/
    F 20H    PROPELLER DIAMETER,26X,5HDP = ,F6.1,7H INCHES/
    F 15H    PROPELLER RPM,30X,6HRPM = ,F6.0,4H RPM/
    F 10H    ALTITUDE,35X,6HALT = ,F6.0,3H FT/)
      PRINT 101,WS,VSO,S,AR,C,EAR,BE,CE,WBE,AD,CDO
101 FORMAT(20H   OUTPUT QUANTITIES://
    F 14H    WING LOADING,31X,6HW/S = ,F6.3,7H LB/FT2/
    F 24H    STALL SPEED WITH FLAPS,21X,6HVSO = ,F6.1,4H MPH/
    F 11H    WING AREA,36X,4HS = ,F6.1,4H FT2/
    F 14H    ASPECT RATIO,32X,5HAR = ,F6.2/
    F 7H     CHORD,40X,4HC = ,F6.2,3H FT/
    F 24H    EFFECTIVE ASPECT RATIO,21X,6HEAR = ,F6.2/
    F 16H    EFFECTIVE SPAN,30X,5HBE = ,F6.2,3H FT/
    F 17H    EFFECTIVE CHORD,29X,5HCE = ,F6.2,3H FT/
    F 24H    EFFECTIVE SPAN LOADING,20X,7HW/BE = ,F6.2,7H LB/FT2/
    F 11H    DRAG AREA,35X,5HAD = ,F6.2,4H FT2/
    F 28H    ZERO-LIFT DRAG COEFFICIENT,17X,6HCDO = ,F6.3)
      PRINT 102,VMINS,THPM,DMIN,RSMIN,LDMAX,CLMINS,RCSTAR,VPROP,TS
    P ,MP
102 FORMAT(
    F 27H    AIRSPEED FOR MINIMUM SINK,16X,8HVMINS = ,F6.1,4H MPH/
    F 41H    MINIMUM POWER REQUIRED FOR LEVEL FLIGHT,3X,7HTHPM = ,F6.2,
    F 3H HP/
    F 14H    MINIMUM DRAG,30X,7HDMIN = ,F6.1,3H LB/
    F 19H    MINIMUM SINK RATE,24X,8HRSMIN = ,F6.1,4H FPM/
    F 28H    MAXIMUM LIFT-TO-DRAG RATIO,15X,8HLDMAX = ,F6.2/
    F 34H    LIFT COEFFICIENT AT MINIMUM SINK,8X,
    F 9HCLMINS = ,F6.2/
    F 26H    MAXIMUM IDEAL CLIMB RATE,16X,9HRCSTAR = ,F6.1,
    F 4H FPM/
    F 37H    REFERENCE PROP AIRSPEED FOR .74 EFF,6X,8HVPROP = ,
    F F6.1,4H MPH/
    F 25H    IDEALIZED STATIC THRUST,21X,5HTS = ,F6.1,3H LB/
    F 27H    PROPELLER TIP MACH NUMBER,19X,5HMP = ,F6.4//
    F 4X,8HAIRSPEED,2X,13HRATE-OF-CLIMB,2X,8HPROP EFF,4X,9HSINK RATE,
```

```
      F 5X,11HREYNOLDS NO/
      F 5X,6HV(MPH),6X,7HRC(FPM),7X,3HETA,8X,7HRS(FPM),5X,
      F 15HRE = RHO*V*C/MU/)
        IMAX=0
        ISTAL=1
        SIG=(1.-ALT/145800.)**4.265
        T=518.7-0.00356*ALT
        RMU=1.
        T1=1./3.
        RCMAX=0.
        V=VS1
        RC=0.
        GO TO 21
20      V=V+DELV
        IF(V.LT.VS1) GO TO 20
21      RC1=RC
        VH=V/VMINS
        RSH=.25*(VH**4+3.)/VH
        RS=RSH*RSMIN
        VT=V/VPROP
        T2=SQRT(1.+0.23271*VT**3)
        ETA=0.92264*VT*((1.+T2)**T1-(T2-1.)**T1)*.85
        RC=RCSTAR*ETA-RS
        IF(RC.GT.RCMAX) RCMAX=RC
        RC2=RC
        IF(RC.LE.0.) GO TO 23
        REC=SIG*V*C*9324./RMU
        PRINT 103,V,RC,ETA,RS,REC
103     FORMAT(4X,F6.1,7X,F6.1,7X,F6.4,7X,F6.1,6X,E10.3)
        IF(ISTAL.EQ.1) GO TO 22
        IF(IMAX.EQ.1) GO TO 24
        GO TO 20
22      CONTINUE
        V=0.
        ISTAL=0
        GO TO 20
23      CONTINUE
        IMAX=1
        V=V-DELV*RC2/(RC2-RC1)
        GO TO 21
24      CONTINUE
        WV2=W*V*V
        FP=RCMAX*WU/33000./BHP*(1.-(VS0/V))
        PRINT 104,FP,WV2
104     FORMAT(
      F /30H  PERFORMANCE RATING PARAMETER,16X,5HFP = ,F6.4/
      F 26H  KINETIC ENERGY PARAMETER,19X,6HWV2 = ,E10.3
      F ,8H LB MPH2/)
        END
```

```
INPUT PARAMETERS:

STALL SPEED WITHOUT FLAPS                    VS1    =    67.0 MPH
MAXIMUM LIFT COEFFICIENT                     CLMAX  =     1.530
MAXIMUM LIFT COEFF WITH FLAPS                CLMAXF =     2.100
GROSS WEIGHT                                    W   =   1500. LB
USEFUL LOAD                                    WU   =    600. LB
WING SPAN                                       B   =   20.83 FT
AIRPLANE EFFICIENCY FACTOR                      E   =    0.744
ENGINE BRAKE HORSEPOWER                        BHP  =    150. HP
MAXIMUM LEVEL FLIGHT SPEED                    VMAX  =   180.0 MPH
PROPELLER DIAMETER                             DP   =    72.0 INCHES
PROPELLER RPM                                  RPM  =   2700. RPM
ALTITUDE                                       ALT  =      0. FT

OUTPUT QUANTITIES:

WING LOADING                                   W/S  =  17.566 LB/FT2
STALL SPEED WITH FLAPS                         VS0  =    57.2 MPH
WING AREA                                       S   =    85.4 FT2
ASPECT RATIO                                   AR   =     5.08
CHORD                                           C   =     4.10 FT
EFFECTIVE ASPECT RATIO                         EAR  =     3.78
EFFECTIVE SPAN                                 BE   =    17.97 FT
EFFECTIVE CHORD                                CE   =     4.75 FT
EFFECTIVE SPAN LOADING                         W/BE =    83.47 LB/FT
DRAG AREA                                      AD   =     3.02 FT2
ZERO-LIFT DRAG COEFFICIENT                     CDO  =    0.0353
AIRSPEED FOR MINIMUM SINK                      VMINS=    78.3 MPH
MINIMUM POWER REQUIRED FOR LEVEL FLIGHT        THPM =    39.42 HP
MINIMUM DRAG                                   DMIN =   163.6 LB
MINIMUM SINK RATE                              RSMIN=   867.3 FPM
MAXIMUM LIFT-TO-DRAG RATIO                     LDMAX=     9.17
LIFT COEFFICIENT AT MINIMUM SINK               CLMINS=    1.12
MAXIMUM IDEAL CLIMB RATE                       RCSTAR= 3300.0 FPM
REFERENCE PROP AIRSPEED FOR .74 EFF            VPROP=    67.4 MPH
IDEALIZED STATIC THRUST                        TS   =   970.4 LB
PROPELLER TIP MACH NUMBER                      MP   =   0.7711

 AIRSPEED   RATE-OF-CLIMB   PROP EFF    SINK RATE      REYNOLDS NO
  V(MPH)       RC(FPM)        ETA       RS(FPM)      RE = RHO*V*C/MU

   67.0        1175.9        0.6278       895.8        0.256E+07
   70.0        1235.7        0.6418       882.4        0.268E+07
   80.0        1384.0        0.6824       867.9        0.306E+07
   90.0        1463.3        0.7147       895.3        0.344E+07
  100.0        1482.0        0.7404       961.3        0.382E+07
  110.0        1445.6        0.7607      1064.7        0.420E+07
  120.0        1357.6        0.7768      1205.7        0.459E+07
  130.0        1220.2        0.7895      1385.2        0.497E+07
  140.0        1034.4        0.7997      1604.6        0.535E+07
  150.0         800.1        0.8078      1865.7        0.573E+07
  160.0         517.0        0.8144      2170.5        0.612E+07
  170.0         183.8        0.8197      2521.3        0.650E+07
  174.8           6.5        0.8219      2705.8        0.668E+07

PERFORMANCE RATING PARAMETER                   FP  = 0.1209
KINETIC ENERGY PARAMETER                       WV2 = 0.458E+08 LB MPH2
```

APPENDIX F. AIRPLANE EFFICIENCY FACTOR, e; GROUND EFFECT

The airplane efficiency factor, e, is used to modify Prandtl's lifting line theory for wings so that the theory can be used for performance calculations for complete airplanes. This is done by multiplying the actual aspect ratio of the airplane wing, AR, by the factor "e" to obtain an effective aspect ratio, eAR, that is then used with the theory (see Part 2). Agreement of results is good as long as the airplane efficiency factor is so chosen that the lift-dependent part of the drag acts as it it were the induced drag of an airfoil with an effective aspect ratio eAR.

If we compare the drag of the ideal airplane with effective aspect ratio eAR to the drag of the airplane with each of the components added to form the total drag, we will see how to calculate the drag area and the airplane efficiency factor. First, for comparison, we write the term $C_D S$, where C_D is the drag coefficient (including induced drag) and S is wing area.

$$C_D S = \left(C_{D,0} + \frac{C_L^2}{\pi eAR} \right) S . \tag{F.1}$$

The term $C_{D,0} S$ is defined as the drag area, A_D. If we add together the contributions to the drag of the airplane for each component, we have

$$C_D S = \underbrace{C_{D,wing} S (1 + K_{wing} C_L^2)}_{\text{parasite drag of the wing}} + \underbrace{C_{D,fuse} S_{fuse} (1 + K_{fuse} \alpha_{deg}^2)}_{\text{parasite drag of the fuselage}}$$

(cross-sectional area of fuselage; angle of attack of fuselage)

$$+ \ldots \underbrace{C_{D,\pi} S_\pi}_{\substack{\text{parasite drag} \\ \text{of other parts}}} + \underbrace{\frac{C_L^2}{\pi AR} (1 + \delta) S}_{\substack{\text{induced drag} \\ \text{of airfoil}}} , \tag{F.2}$$

where the term K_{wing} gives the change in the parasite drag versus lift coefficient as determined from the airfoil section data (see Appendix H). The angular-dependent drag coefficient for the fuselage is referred to the cross-sectional area of the fuselage, S_{fuse}. The term $C_{D,\pi} S_\pi$ is meant to represent the component drag of the other parts of the airplane, such as the drag of struts, landing gear, antennas, tail surfaces, etc. And, the induced drag of the airfoil includes the theoretical correction of the planform in the term $(1 + \delta)$. For an elliptical planform Prandtls lifting line theory tells us that δ will be equal to zero.

If we compare the two expressions (F.1) and (F.2) at zero lift conditions ($\alpha = 0$, $C_L = 0$), we see that the drag area is given by

$$A_D = C_{D,wing} S + C_{D,fuse} S_{fuse} + C_{D,\pi} S_\pi \ldots \tag{F.5}$$

This expression is discussed in Appendix G where the drag area is estimated by adding the drag areas for the airplane components.

To compare the lift-dependent parts of the two expressions, we first need to determine the lift-slope with corrections for finite aspect ratios. The change with aspect ratio of the slope of the curve of the lift coefficient, C_L, versus angle-of-attack, α, is given by Wood (1963),

$$\frac{dC_L}{d\alpha_{deg}} = 0.110 \frac{AR}{AR + 3} \quad (1/deg). \tag{F.6}$$

Therefore, if we take a linear approximation for the variation of lift coefficient

$$C_L = \frac{dC_L}{d\alpha_{deg}} \alpha_{deg} \quad , \tag{F.7}$$

we can solve for α_{deg} and substitute into (F.2) and equate the C_L^2 dependent parts with the induced drag part of (F.1):

$$\frac{1}{e} = \underbrace{(1 + \delta) + \pi AR\, C_{D,wing} K_{wing}}_{\frac{1}{e_w}:\ \text{wing efficiency factor}} + \underbrace{\pi C_{D,fuse} K_{fuse} \frac{(AR + 3)^2}{.012\, AR} \frac{S_{fuse}}{S}}_{\frac{1}{e_{fuse}}:\ \text{fuselage correction}} \tag{F.8}$$

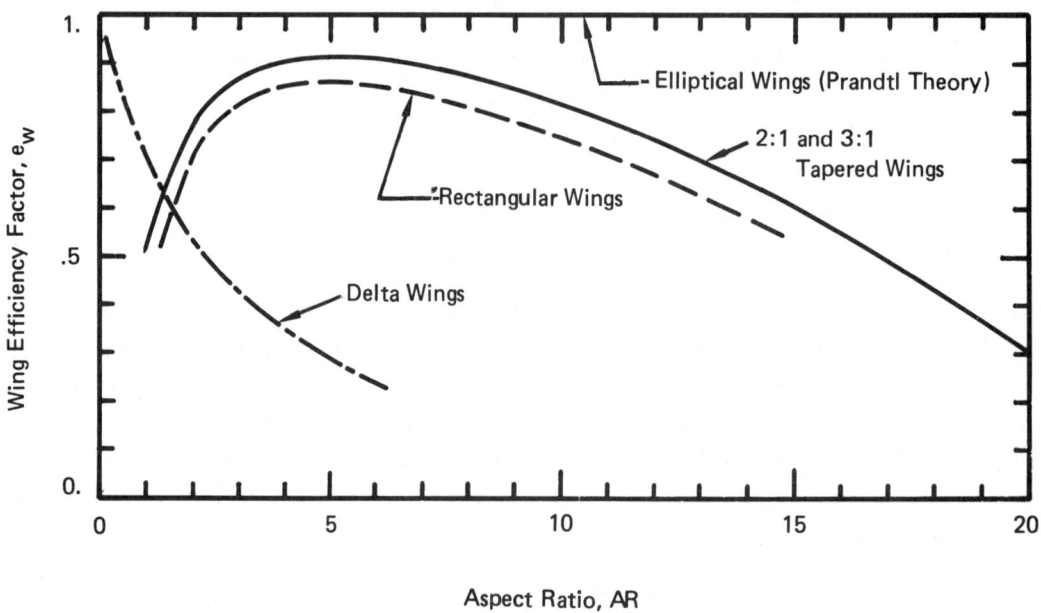

Figure F.1. Variation of Wing Efficiency Factor with Aspect Ratio.

The first three terms on the right hand side are grouped together in terms of the wing efficiency factor. This factor, e_w, is plotted as a function of aspect ratio for various wing planforms by Wood (1963), (Figure F.1) resulting in the recommended-practice line for the various wing planforms. However, if the airfoil section that we choose is a modern low-drag airfoil, these curves may not be very accurate, since the drag coefficient may have a "drag bucket" (as described in Appendix H) and the K_{wing} term describing the shape of C_D vs C_L curve will be small. This will cause a considerable error for airplanes with high aspect ratio wings, such as sailplanes. For an estimate of the airplane efficiency factor for sailplanes, see the paper by Martin (1977). The value for the wing efficiency factor for delta wings is estimated from data in Appendix E of the book by Nicolai (1975).

The last term on the right hand side of (F.8), used to correct for the effects of the drag dependence of the fuselage on angle of attack, is plotted in Figure F.2 for round and square fuselages in terms of $\frac{\Delta(1/e)_{fuse}}{S_{fuse}/S}$ versus aspect ratio, AR (from Wood, 1963).

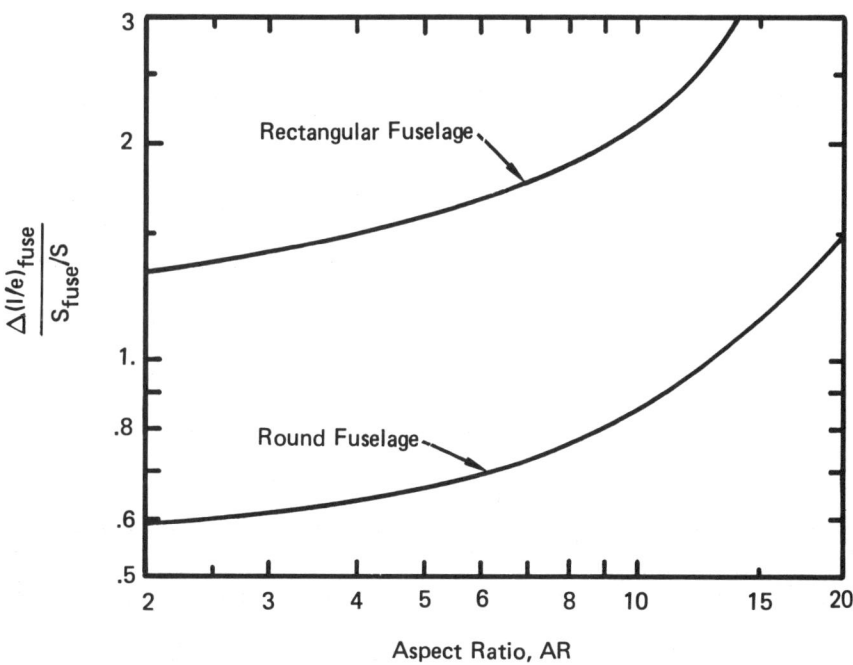

Figure F.2. Effect of Fuselage on Airplane Efficiency Factor.

The airplane efficiency factor for the T-18 can be calculated from the Figures F.1 and F.2 and a knowledge of the size of the components of the aircraft. Following the calculations of Henderson and Roemer (1977), we first use the aspect ratio of the T-18 (AR = 5) to find the value of the wing efficiency factor, e_w. From Figure F.1 for the wing efficiency factor for a rectangular wing with an aspect ratio of 5 is 0.85, so that $1/e_w = 1.176$. The additional term for the effect of the fuselage is determined from Figure F.2, together with the wing area and an estimate of the frontal area of the fuselage, S_{fuse}. For an aspect ratio of 5 and for a rectangular fuselage, we find from Figure F.2 that

$$\frac{\Delta(1/e)_{fuse}}{S_{fuse}/S} = 1.6$$

If the fuselage area is estimated at about 3 x 3 square feet (9 ft²), then

$$\Delta(1/e)_{fuse} = 1.6 \times \frac{9}{86} = 0.167$$

Therefore,

$$(1/e) = (1/e_w) + \Delta(1/e)_{fuse} = 1.176 + 0.167 = 1.343.$$

Thus, the airplane efficiency factor is $e = \left(\frac{1}{1.343}\right) = 0.744$. This factor is used in the sample calculation for the effective span, b_e, using relation ③.

Ground-effect changes the effective airplane efficiency factor. When the airplane is flying within one wing span of the surface, the induced drag of the wings is greatly reduced. This result is due to the "image system" that potential flow theory uses to make the ground plane into a streamline. In this case, the vortex system from the image wing interacts with the vortex system of the wing itself to reduce the particle displacements as the air flows over the wing. The effect is to greatly enhance the airplane efficiency factor according to Figure F.3. Wing efficiency is multiplied by the factor k_{gd}, which is a function of the height above the ground divided by the wingspan, h/b.

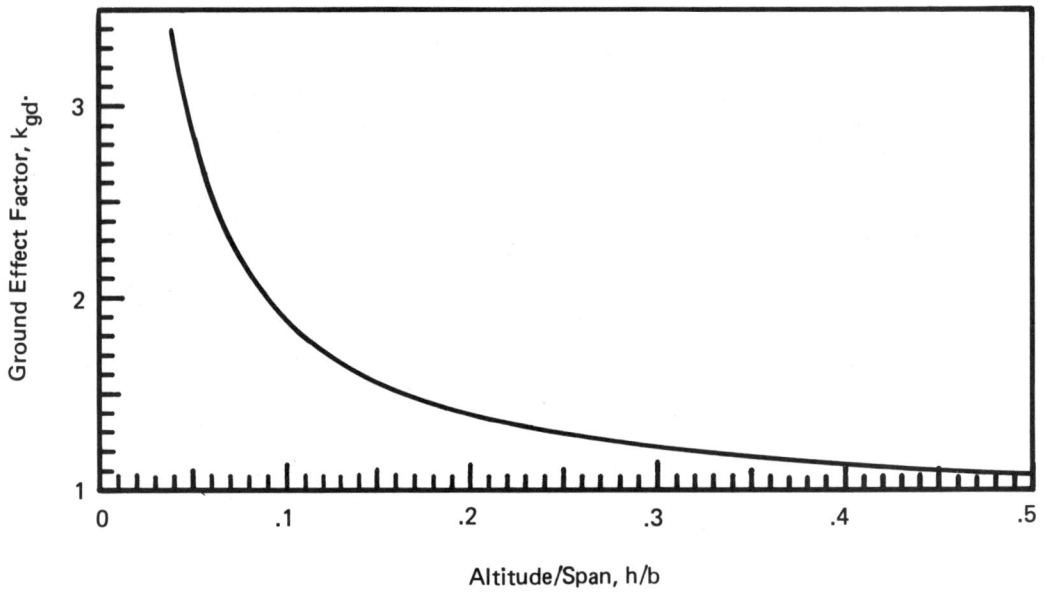

Figure F.3. Altitude Dependence of Ground Effect Factor.

APPENDIX G. DRAG ANALYSIS

The art of estimating the drag of a complete airplane is a skill that can be developed by study and practice, practice, practice. Since the topic is broad enough to be dealt with in a book by itself — for instance, **Fluid Dynamic Drag** by Dr. S. Hoerner (1965) — the discussion presented here can only be considered as an introduction to the subject. For preliminary analyses, the drag of the airplane can be estimated by considering the drag of the major components of the airplane. If we calculate the drag area of each component and add them together with additional factors to account for interference effects, we can use this drag area in the performance analysis discussed in Part 1 of the text. In the preliminary stages the size has not been completely specified, so the drag area that is calculated the first time through may not be accurate enough for the final performance calculation. For example, we need to find the contribution of the fuselage and the horizontal and vertical tail surfaces to the drag area, but the dimensions of these components cannot be completely determined until we make an analysis to find the moment arms and tail areas that will satisfy the stability requirements. Since the templates that are used to calculate the rate-of-climb performance can be easily used to find the effect of a change in the drag area, it is only necessary at the beginning stages to make as good a first estimate of the drag as we can, and then refine the estimate as the design progresses to a more finalized form.

As a first approximation the drag area can be split into two terms: one due to pressure drag (caused by the separation of the boundary layer resulting in a net pressure difference between the front and rear halves of the airplane) and the viscous drag (directly related to the shear stresses caused by the flow of the viscous air over the surface of the airplane). In the pressure drag term, the drag area will be proportional to the frontal area of the airplane — the projected area that would be seen by an observer at the front of the airplane. The skin friction drag term will be proportional to the total wetted surface area — the surface of the airplane that would get wet if the air surrounding the airplane were water. We can assume for a first approximation that the drag area, A_D, can be represented by

$$A_D = C_{D,f} A_f + C_{D,w} S_w \tag{G.1}$$

where $C_{D,f}$ is the drag coefficient for the pressure drag, $C_{D,w}$ is the drag coefficient for the skin friction drag, A_f is the frontal area, and S_w is the wetted area, respectively. If we take values of 0.04 and 0.003 for $C_{D,f}$ and $C_{D,w}$, we can get reasonable first estimates for the drag area of a fairly streamlined airplane. (Remember to use the correct reference areas, A_f and S_w, the frontal and wetted areas, for this calculation.) The wetted area will be twice the wing planform area plus the surface area of the fuselage and tail. This approximate analysis should only be used in the very early design stages.

For a better estimate of the drag area, we can add the drag areas of the various components, where the individual drag coefficients of the components, $C_{D,\pi}$, are determined from data or books such as the one by Hoerner. The individual drag coefficients are based on various reference areas, S_π, which depend on the component being described. For instance, the drag coefficient for the wing is usually based on the planform area of the wing. The drag coefficient for the fuselage, on the other hand, may be based on the wetted area or the frontal area. It is important, therefore, to know which reference area is being used in the definition of the drag coefficient so that we can calculate the drag area correctly.

FLAT PLATE DRAG

The lowest possible drag for a flat-slided airplane component is that for the laminar flow over a flat plate aligned with the flow velocity vector. In laminar flow, the Reynolds number is so low that the viscous forces dominate the inertia forces and there is a smooth flow of air over the plate. (See Appendix I for the definition and a nomogram for the Reynolds number). For laminar flow over a plate, the drag coefficient based on the surface (wetted) area is given by the Blasius formula

$$C_{D,w} = \frac{1.328}{\sqrt{Re_\ell}} \quad , \text{(laminar)} \tag{G.2}$$

where the Reynolds number, Re_ℓ, is based on the length, ℓ, of the flat plate (or the chord length if we are considering the drag of the wing). However, small disturbances in the boundary layer — the thin region of air that is slowed down next to the surface — cause the flow to become unstable and to break down into turbulent flow where there is rapid mixing and higher drag. For turbulent flow, the drag coefficient based on the wetted area of the flat plate is related to the Reynolds number by an approximation to the data of Schoenherr (see Hoerner, 1965)

$$C_{D,w} = \frac{1}{(3.46 \log_{10} Re_\ell - 5.6)^2} \quad . \text{(turbulent)} \tag{G.3}$$

If the flow starts to become turbulent when the Reynolds number reaches a critical value, the flow is said to be transitional. For the flat plate the transitional Reynolds number is usually in the range of 300,000 to 700,000. In the transition regime, the drag coefficient can be taken to be a curve fit of the form

$$C_{D,w} = C_{D,w}(\text{turb}) - 1700/Re_\ell \quad \text{(transitional)} \tag{G.4}$$

where the constant 1700 is chosen to match the transitional data.

The drag coefficients for laminar, turbulent and transitional boundary layer flow over a flat plate (equations G.2, G.3 and G.4) are plotted in Figure G.1. These drag coefficients are based on the wetted area and can be considered as lower bounds for the drag of any flat-sided aircraft component. In the case of aircraft fuselages, we can first calculate the Reynolds number based on the fuselage length, find the drag coefficient from Figure G.1, and then multiply by the fuselage wetted area to obtain the drag area. For the wings, we can do better, because we will be choosing a particular airfoil (see Appendix H) and we can find the zero-lift drag coefficient from the airfoil data. Otherwise we can use the curves in Figure G.1 to estimate the drag area for our wing. Similarly, for the tail surfaces, we can use the data for the particular airfoil section that we choose for those components.

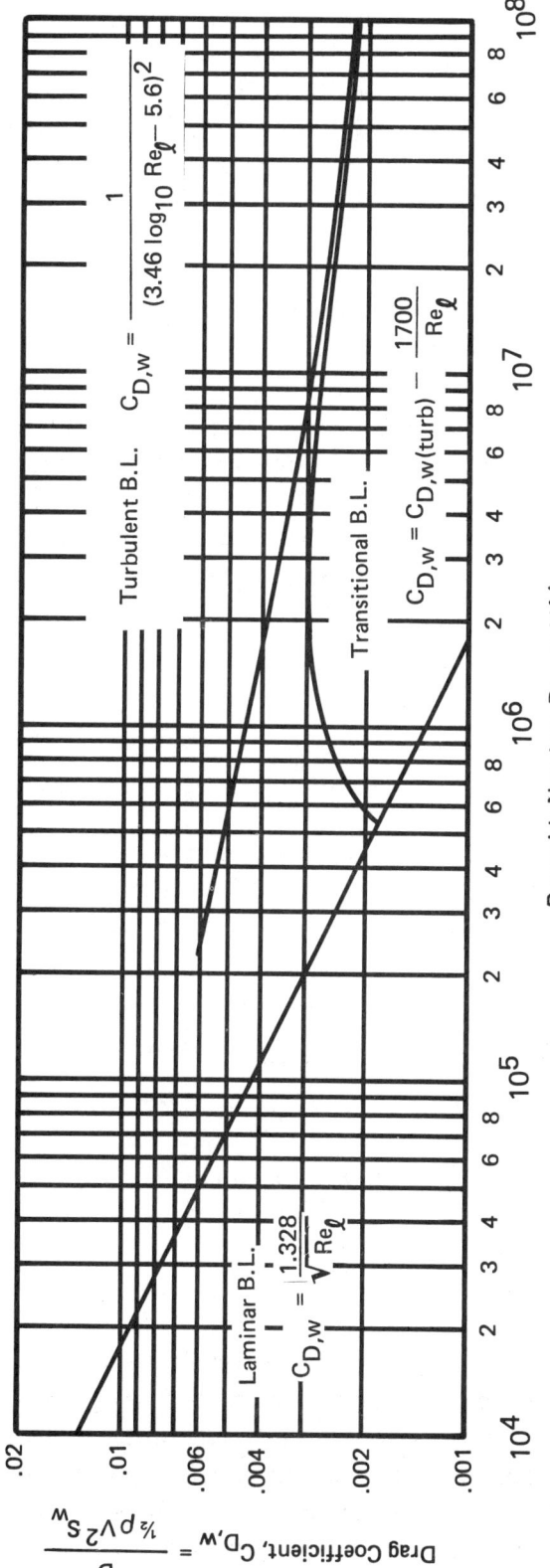

Figure G.1. Drag Coefficient Based on Wetted Area for Flat Plates in Laminar, Turbulent, and Transitional Flow.

WIRES AND CIRCULAR STRUTS; STREAMLINED STRUTS

The drag of wires and struts can sometimes contribute a considerable fraction to the total drag of an airplane. The mechanism for the drag of a circular cylinder (strut or wire) is different from that of the flat plate. For the flat plate, the skin friction at the surface is integrated to give the drag. For blunt bodies, it is the pressure drag that gives the major contribution to the drag. In potential flow theory, it can be shown that the pressure forces are equal on the front and back of a body in an ideal fluid without viscosity. This means that the drag is identically zero — d'Alemberts paradox — since there is no difference in pressure (and since viscous forces are not considered). Air, however, has a small value of viscosity, which causes a considerable change in the flow pattern on the back side of the cylinder. There, the flow separates, leaving a low pressure region directly opposite the high pressure region on the front side of the cylinder, causing a large value for the pressure drag. Theory cannot predict the shape of the drag coefficient for cylinders as a function of the Reynolds number except for very small values. The drag coefficient for a cylinder as a function of the Reynolds number is plotted in Figure G.2 based on the data collected by Hoerner (1965). The Reynolds number, Re_D, is based on the diameter, D, of the cylinder. Therefore, to obtain the drag area for a strut or bracing wires, calculate the Reynolds number based on the diameter (see Appendix I), find the drag coefficient for a cylinder at this Reynolds number from the curve in Figure G.2, and multiply by the length and the diameter of the tube or wire.

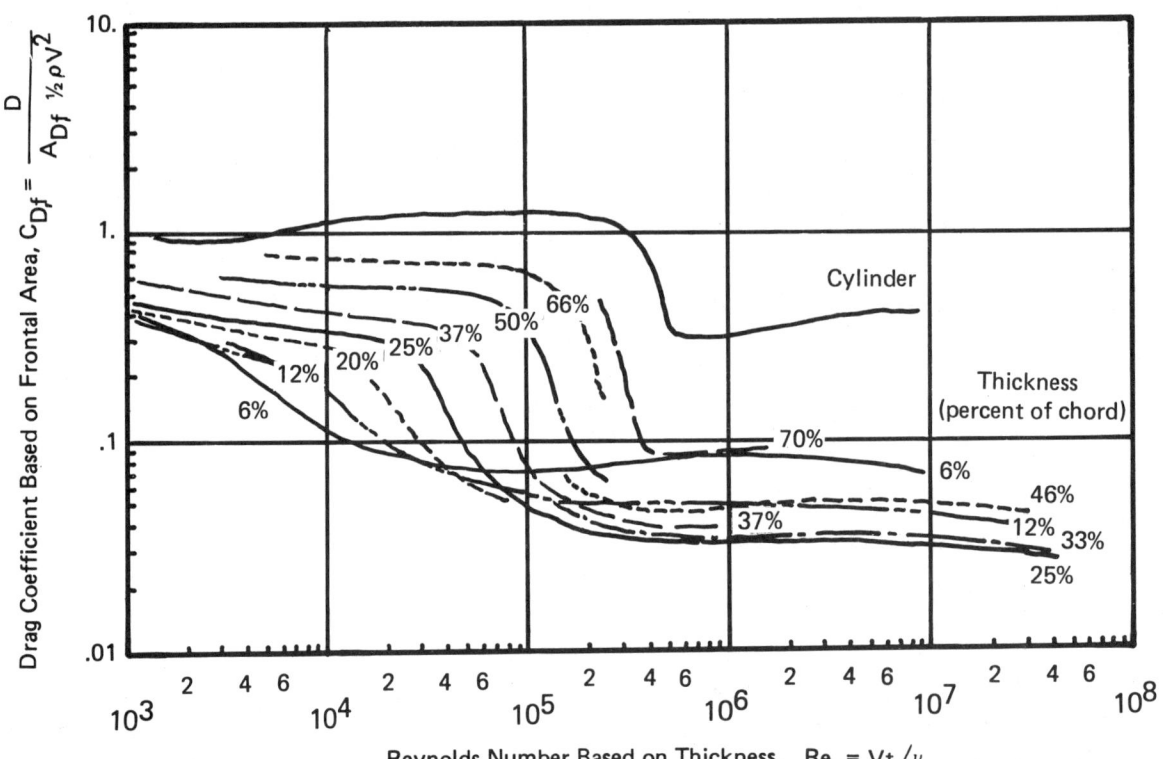

Figure G.2. Drag of Streamlined Struts.

For cylinders, there is a critical Reynolds number, $Re_{cr} \approx 6 \times 10^5$, above which the drag coefficient reduces to a value about one-third the sub-critical value. The high drag of the sub-critical flow is caused by the large separated region on the back side of the cylinder as described before. However, at the higher Reynolds number, the boundary layer on the cylinder becomes turbulent closer to the leading edge stagnation point. It is this turbulent boundary layer that is better able to negotiate the adverse pressure rise on the back of the cylinder. Since the turbulent boundary layer can traverse further into the pressure gradient, the boundary layer on the cylinder does not separate as soon as the old laminar boundary layer, and the extent of the separated flow on the rear of the cyliner is smaller. This smaller area of low pressure causes the sharp decrease in drag for cylinders with turbulent boundary layers. We can use this to advantage to decrease the drag of a cylinder if we glue sand onto the front side of the cylinder to act as a boundary layer "trip" to cause a turbulent boundary layer. The same principle holds for golf balls, where the dimples cause a turbulent boundary layer and lower drag.

The drag of streamlined struts and flying wires can be much lower than the drag of circular struts and wires with the same cross-sectional thickness. The characteristics of the variation of the drag coefficient with respect to Reynolds number is similar to that of the circular cylinder and decreases as can be seen in Figure G.2. The drag coefficient decreases with increasing Reynolds number until a critical Reynolds number is reached. Then, the drag coefficient dramatically decreases as the turbulent boundary layer on the surface finds it easier to negotiate the pressure increase on the backside of the streamlined section. This prevents the large area of laminar separation and the associated high drag. The data presented in the book by Hoerner has been replotted in Figure G.2 so that the drag coefficient is based on the frontal area of the strut. This way we can see exactly how much drag we can save by streamlining the flying wires or struts. For instance, at a Reynolds number based on thickness of 8×10^5 (roughly corresponding to a 1 inch strut at 100 mph), we can decrease the drag by an order of magnitude — from 0.35 to 0.035 — by adding a streamlined fairing with a thickness-to-chord ratio of 0.25. Alternatively, we see that a streamlined section can be ten times larger than a circular wire and still have the same drag. Note that the Reynolds number in the figure is based on the thickness of the strut or wire. At each Reynolds number an optimum thickness ratio will give the least drag. This yields the value for the chord that should be chosen for a strut with a given cross-section.

DRAG AREA FOR A HANG GLIDER PILOT

The drag area for an average man in various positions is reported in the book by Hoerner (1965). In an upright position relative to the wind the drag area is found to be about 9 ft^2; in a seated position, as in my Seagull V hang glider, the drag area is found to be about 6 ft^2; and for a prone position (with arms at the side) the drag area is about 1.2 ft^2. These numbers, together with the values for the drag coefficients for the boundary layer on the sail, wires, and circular struts can be used to give a first approximation to the drag area for a hang glider where the pilot, wires, struts and sailcloth are all hanging out in the breeze together. For the sample calculation of the powered Quicksilver, it was assumed that the drag area was about 16 ft^2 (6 ft^2 for the pilot and 10 ft^2 for the hang glider).

FUSELAGE DRAG

The drag of a fuselage is mainly caused by the pressure difference caused by the separation of the boundary layer. To reduce the drag we would like to keep the frontal area of the fuselage to a minimum. Once we have made the frontal area as small as possible, we need to choose the fineness ratio of the fuselage to streamline the structure as much as possible. The choice of the fuselage shape is often dictated by other considerations, such as the length of the structure to support the horizontal and vertical tail surfaces.

If we are free to choose the optimum fineness ratio (length-to-diameter ratio, ℓ/d), we can use the data plotted in Figure G.3. This information has been obtained from Hoerner, but has been replotted in terms of the drag coefficient based on frontal area and the Reynolds number based on the maximum diameter of the fuselage. We have assumed that the wetted surface area is related to the frontal area by the following expression

$$\frac{S_w}{A_f} = 3\frac{\ell}{d},$$

which is a good approximation for normal streamlined shapes. We see from the figure that there is a different optimal shape for each Reynolds number regime. At low values of the Reynolds number, a fineness ratio of 8 gives the minimum drag; at high Reynolds numbers, the optimal fineness ratio is about 3; and at intermediate values, a fineness ratio of 5 gives the least drag.

To find the drag area of the fuselage, calculate the Reynolds number based on the maximum thickness, Re_d (see Appendix I). Then, find the drag coefficient for the chosen value of the fineness ratio, ℓ/d, from Figure G.3. Finally, multiply the drag coefficient by the cross-sectional area to obtain the drag area.

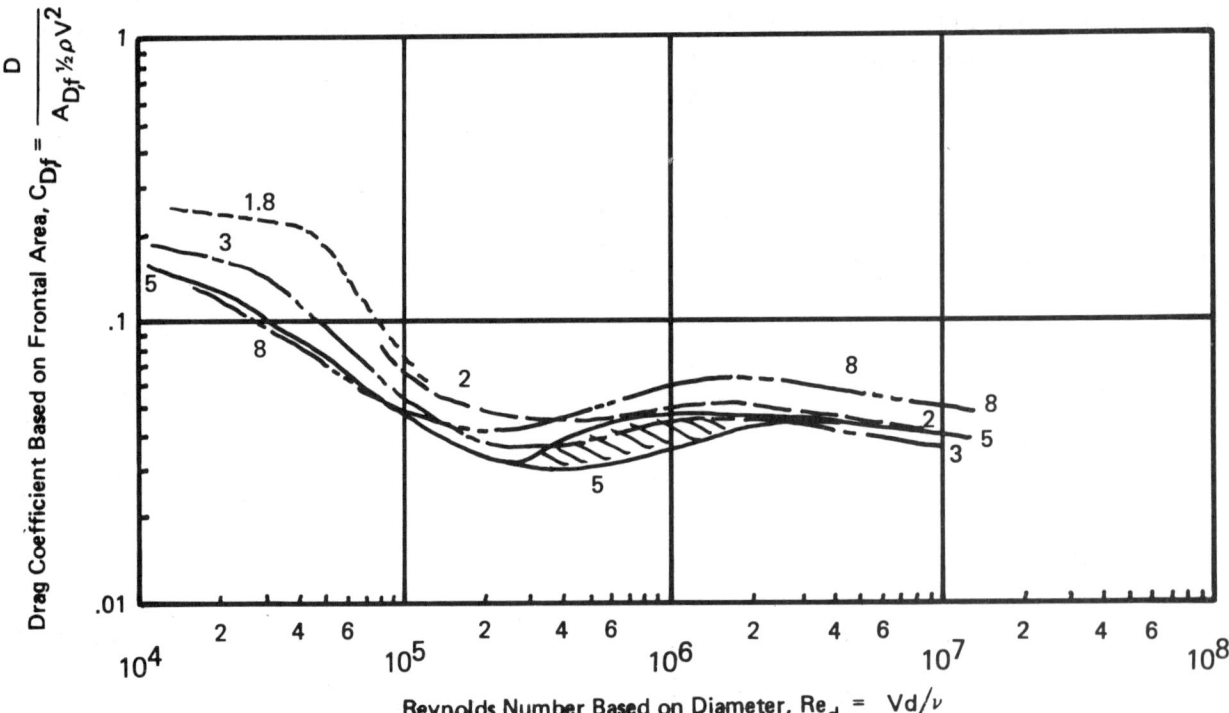

Figure G.3. Drag of Fuselage Shapes.

DRAG OF AIRCRAFT COMPONENTS

The drag of various aircraft components is discussed throughly by Hoerner (1965) and the reader is advised to consult that source. In that book are representative values for the drag of fuselages, with various canopies; engine nacelles, radiator installations; floats and boat hulls; flaps and dive brakes; landing gears; external loads; windmilling and stopped propellers; and a multitude of other practical examples. As an example, the drag of the landing gear can be estimated from the following values from the drag coefficient based on the wheel frontal area (diameter x tire width)

Nose wheel and strut	0.5 – 0.8	
Main gear with well faired wheel pants	0.15 – 0.3	$C_{D,f}$
Main gear with wheels and struts exposed	0.3 – 0.5	

The cooling drag is complicated and depends upon the ratio of the frontal area to the area around the cylinders, the pressure drop between the intake and the space behind the baffled cylinders, and the flow velocity between the cylinder fins. The cooling drag can contribute a significant amount of drag area if we are trying to obtain a fast clean airplane. For the further details of the analysis of this and other difficult drag-related problems, see Hoerner!

When all of the drag areas of the airplane are added, we should add another 10 percent to account for additional interference drag, because the drag of the combined parts is usually more than the drag of each component considered separately.

APPENDIX H. AIRFOIL SELECTION

Several decisions must be made in the selection of an airfoil for a design. First, the wing should be thick enough that it can be built to withstand the aerodynamic loads that will be applied during flight. This means that thicker wings will be more appropriate if we want to build a cantilever wing, because we can build a deeper spar. A thin wing can be built, but it will have an increased weight. A thick wing, however, usually has a larger value for the minimum drag coefficient. If we want to keep the drag of the airplane to a minimum, we should choose a thin profile that has a low drag coefficient.

One problem with thin profiles is that they usually have bad stall characteristics. That is, the shape of the lift coefficient versus angle of attack curve (C_L vs α) is too sharply curved at the maximum. A sharp curve at the stall condition will be sharply felt in the cockpit when the airplane stalls. A smoothly rounded shape will have a good stall characteristics because the airplane will start to mush as the stall is approached, giving a warning that a stall is imminent. Of course, if we are designing an aerobatic airplane, we would look for a sharply breaking stall in order to perform some of the aerial snap maneuvers. For racing airplanes we would not be that interested in the characteristics near stall, so thin airfoils would be appropriate.

We also have to consider the value for the maximum lift coefficient, $C_{L,max}$. For a wing without flaps, the maximum lift coefficients are greatest for wings which have the thickness-to-chord ratio of 0.12 (12 percent thickness airfoils). However, if flaps are to be used, we find from the NACA data reported by Abbott and Von Doenhoff (1949) that the thick airfoils (18 to 20 percent) will have the highest values of the maximum lift coefficient.

To reduce the trim drag and the torsion loads on the wing, we should look for an airfoil that has a moment coefficient near zero. Trim drag is caused by the tail when it has to overcome the pitching moment of the wing to make the airplane stable. Also, the large torsion loads on the wing caused by the large moments will lead to heavier construction.

The airfoil sections developed by NACA use a numbering system that can be used to estimate the properties of the section by interpretation of the numerical designator. In our example, the T-18 uses the NACA 63_1-412 airfoil. The 6 is the series designation — a family of low drag airfoils with thickness distributions designed to give low drag and high lift characteristics. The 3 denotes the chordwise position of the minimum pressure in tenths of the chord measured from the leading edge for the basic symmetrical section at zero lift. This can be seen in Figure H.1 where $(v/V)^2$ peaks at $x/c = 0.3$, since the variation of $(v/V)^2$ is proportional to the negative of the pressure coefficient. This follows from Bernoulli's equation.

$$p + \frac{1}{2}\rho v^2 = p_\infty + \frac{1}{2}\rho V^2 \quad , \tag{H.1}$$

to give

$$\frac{p - p_\infty}{\frac{1}{2} V^2} = 1 - \left(\frac{v}{V}\right)^2 \quad . \tag{H.2}$$

The pressure distribution affects the boundary layer growth on the airfoil.

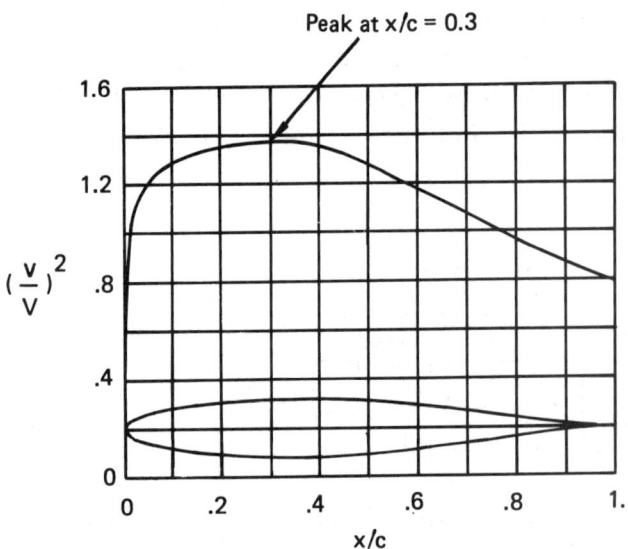

Figure H.1. Velocity Distribution on the Basic Thickness Form, NACA 63_1-012, at Zero Angle of Attack.

Returning to the airfoil designation, the subscript 1 represents the extent of the "drag bucket" around the design lift coefficient. In this case, the low drag region extends plus and minus 0.1 around the design lift coefficient, 0.4, represented by the number 4 in the airfoil designation. The drag bucket can be seen in Figure H.2 where the drag coefficient is plotted against the lift coefficient. The last two digits indicate the maximum thickness of the wing relative to the chord. In this case, the thickness is 12 percent of the chord.

Details of the numbering system for other NACA series sections are described in Abbot and Von Doenhoff (1949). New airfoils are currently being developed at NASA, notably the GA(W)-1, GA(W)-2, and the GA(PC)-1 which have good performance in both low speed and high speed cruise flight. More information about these airfoils can be found in the references. Additionally, Dr. Wortmann and Dr. Liebeck have been actively developing new airfoil sections for sailplanes and for high-lift applications. The airfoil design procedure starts with a given pressure distribution which will give favorable conditions for the development of the boundary layer. Then, an inverse technique is used to find the airfoil that corresponds to this pressure distribution.

To interpret and evaluate the data describing the aerodynamic characteristics of wing sections, we have to know which sets of data are applicable to our specific case. We notice three different sets of Reynolds numbers in Figures H.2 and H.3: 3.0×10^6, 6.0×10^6, and 9.0×10^6; a standard roughness curve at Reynolds number 6 million and two curves for a 0.20c simulated split flap deflected 60 degrees. Therefore, we first have to estimate the Reynolds number for the stall and cruise design conditions. This can be done using the information on Appendix I, where we have a nomogram to find the Reynolds number given the airspeed (mph) and the chord length of the airfoil. Since we do not yet have a chord length, we have to make a first guess, and then modify it if the guess is too far wrong. Assuming a chord length of four feet and a stalling speed of 67 mph, we find from Appendix I that the Reynolds number is about 2.5 million. That means that we should use the curve for Re = 3.0×10^6, in the stall region where the lift coefficient is large since this is the closest value for Reynolds number of the data. At higher speeds, say 180 mph, the

Reynolds number is about 6.5 Million so we will have to assume that our conditions will lie between the data for Reynolds number of 6 and 9 million. The low drag data with the drag bucket in Figure H.2 Corresponds to wings that are nearly perfect, as far as the condition of the surface is concerned. If bugs or dirt build up on the surface, the drag rises to the standard roughness curve — a considerable increase.

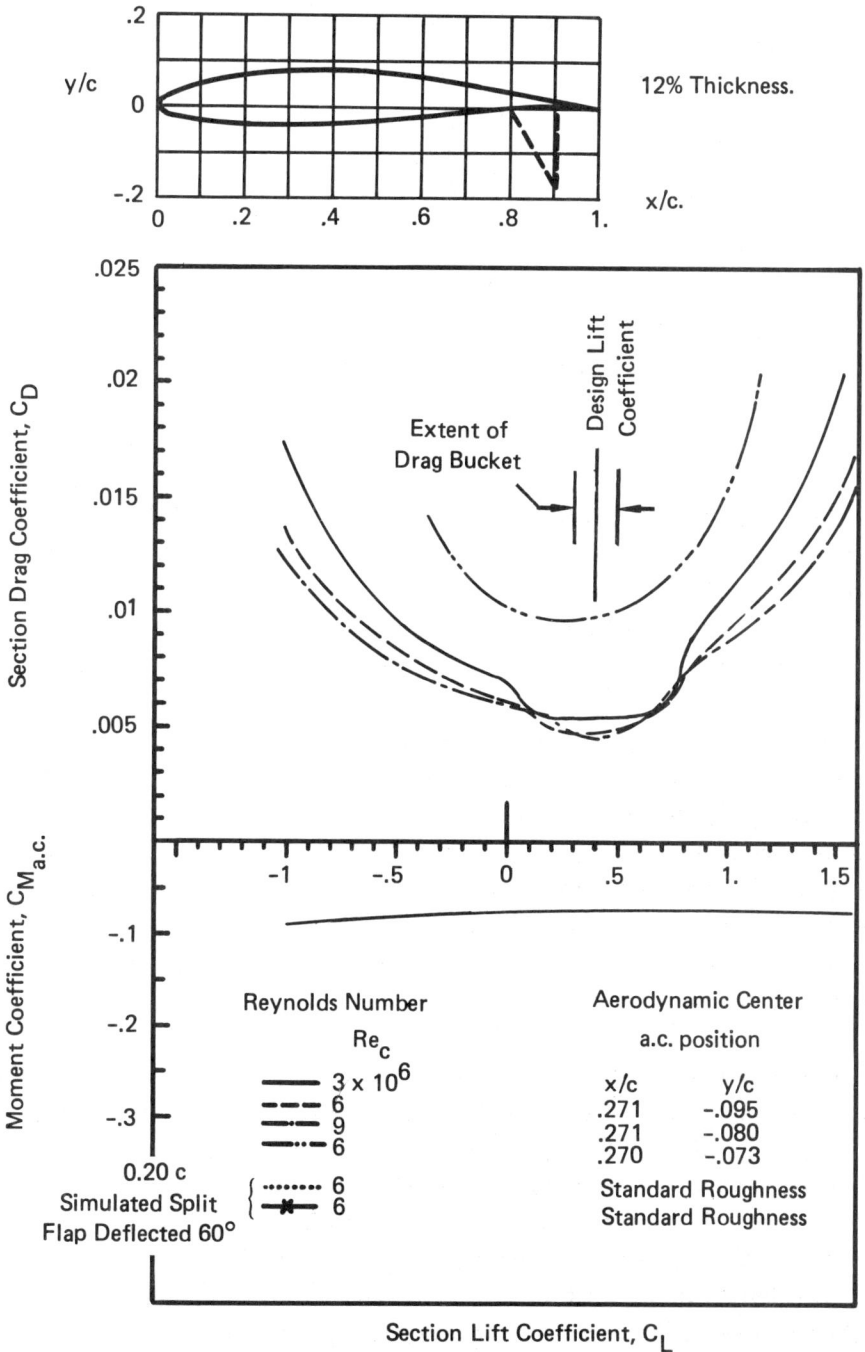

Figure H.2. Lift Dependence of Section Drag Coefficient and Moment Coefficient about the Aerodynamic Center for the NACA 63_1-412 Wing Section

Assuming a smooth surface, let us find the important data from the figures. First, we find from Figure H.2 that the minimum section drag coefficient is about 0.005 at values of the lift coefficient between 0.1 to 0.6. If we multiply this values times the wing area (86 ft^2 for the T-18) we find that the drag area associated with the wing is about 0.43 square feet. If we have a poor surface condition and the drag coefficient increases to 0.1, the drag area will also double to become 0.86 square feet.

From Figure H.2, we find that the moment coefficient about the aerodynamic center is −0.075, and the position of the aerodynamic center is 0.271. These values are important quantities to be used in the stability analysis to determine the size and location of the horizontal tail surfaces.

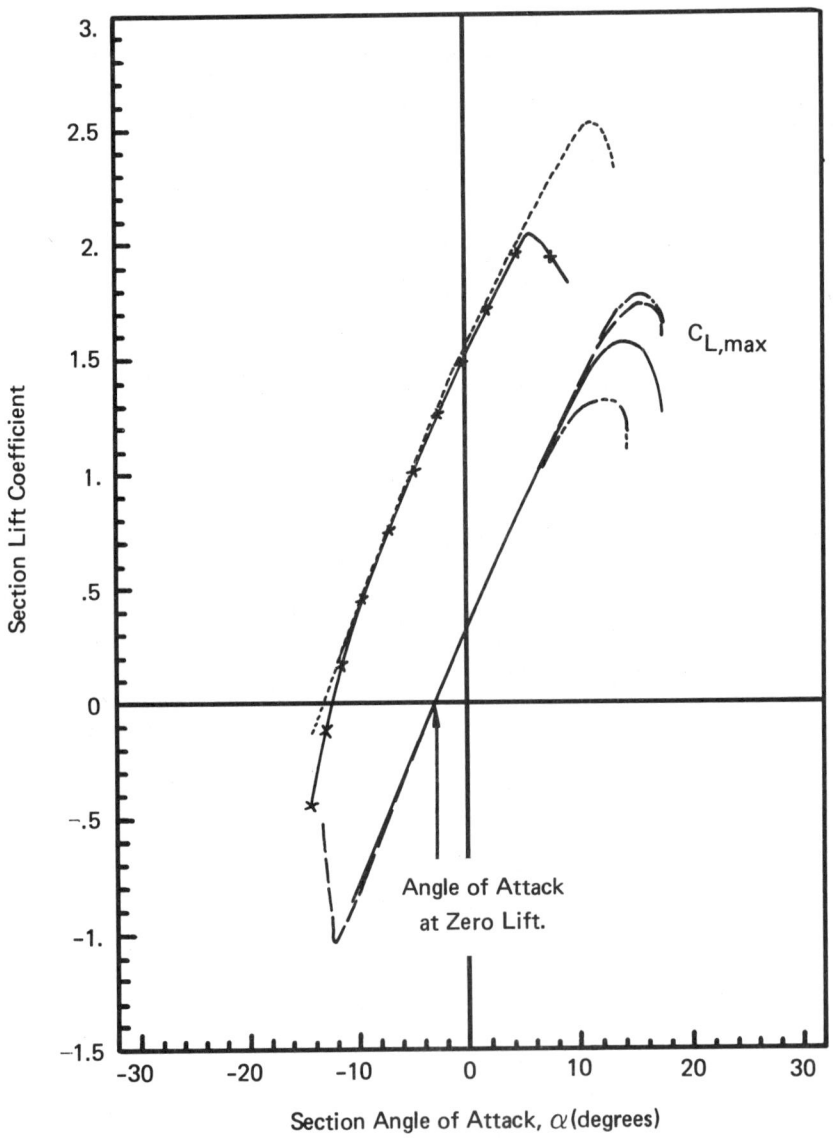

Figure H.3. Angle of Attack Dependence of the Section Lift Coefficient for the NACA 63_1-412 Wing Section. (See Figure H.2 for Legend).

From Figure H.3, we find that the maximum lift coefficient is about 1.52. This will be used in the stall analysis to determine the wing loading that will let us satisfy our desired stall speed. Next, the slope of the curve of lift coefficient versus angle of attack is used in the stability analysis and in the gust loading analysis (corrected for aspect ratio effects). From the data, we see that the lift coefficient increases from zero at $\alpha = -3$ degrees to 1.1 at $\alpha = 7$ degrees. This gives a lift-slope value of 0.11 per degree — in agreement with the theoretical value of 0.110 per degree (or 2π per radian). The angle of attack for zero lift $\alpha_o = -3$ degrees is also important since it will be used to position the wing relative to the fuselage to reduce the drag of the fuselage in flight. The angle of attack at stall is about 14 degrees — another consideration when positioning the wing relative to the fuselage is to find the airplane geometry in the landing configuration.

The application of flaps increases the maximum lift coefficient appreciably, from 1.5 to about 2.5. This will allow us to have either a smaller wing, which will reduce our drag area, or land at a lower approach speed. Notice that the pitching moment is increased substantially, so that we will have to increase the tail moment by adding "up" elevator to maintain equilibrium conditions. The angle of attack for zero-lift is changed to −13 degrees with the application of 60 degrees of flaps. These effects will have to be taken into account when analyzing the requirements for control surface deflections.

APPDENDIX M. HOW TO CALCULATE DRAG AREA, A_D, AND AIRPLANE EFFICIENCY FACTOR, e, FROM FLIGHT TEST DATA.

As described in Part 2 (starting on page 120), the present performance analysis is based on a parabolic fit to the drag polar for the whole airplane. That is, the drag coefficient, C_D, can be represented by

$$C_D = C_{D,O} + \frac{C_L^2}{\pi e AR}$$

where $C_{D,O}$, the zero-lift drag coefficient, and e, the airplane efficiency factor, are the two parameters to be determined from theoretical analyses (Appendices F and G) or from flight test data.

If the other constants for the airplane are known (W, the gross weight, b, the wing span, and σ, the density of air relative to the standard sea level density), then two separate tests will enable us to find the drag area ($A_D = C_{D,O} S$) and the airplane efficiency factor, e. First, record the power-off sink rate as a function of airspeed, making sure that the glide angle is less than about 20 degrees so that the small angle assumption is not violated. Then, record the thrust horsepower required for various level flight speeds, which requires that the engine power output divided by the propeller efficiency be known for the manifold pressure, engine RPM, propeller RPM, and flight speed. Charts for this purpose are usually available from the engine and propeller manufacturers.

Then, plot the sink rate, R_S, versus airspeed, V, on the 2x2 logarithmic graph paper described on page 75. Then, draw a "best-fit" curve through the data with the plastic sink-rate template. From the graphically determined reference sink rate, $R_{S,min}$, and speed for minimum sink, V_{minS}, the drag area and efficiency factor are given by

$$A_D = 1.1108 \frac{W R_{S,min}}{\sigma V_{minS}^3} \quad \text{and} \quad e = 14603 \frac{W}{\sigma b^2 V_{minS} R_{S,min}}$$

This value for the drag area will include the effects of the drag of the windmilling propeller.

The corresponding "best-fit" curve for logarithmic plot of THP versus V will locate the minimum required thrust horsepower, THP_{min}, and V_{minS}, the speed for minimum sink rate (or minimum required power). Then, A_D and e are given by

$$A_D = 36656.3 \frac{THP_{min}}{\sigma V_{minS}^3} \quad \text{and} \quad e = .4425 \frac{(W/b)^2}{\sigma THP_{min} V_{minS}}$$

These expressions are basically restatements of relation ⑦ on the nomogram.

An alternative graphical method, as discussed by Kohlman (1979), can be used to calculate these factors. Plot y (=V THP or V W R_S/33000) versus x (= V^4) on linear paper. Then determine the slope, A, and the intercept, B, for the best straight-line fit of the data

$$y = Ax + B.$$

Then, the drag area and the efficiency factor are given by

$$A_D = 146625 \, A/\sigma \quad \text{and} \quad e = 1.0427 \frac{(W/b)^2}{\sigma B}$$

APPENDIX I. REYNOLDS NUMBER, Re = $\rho V \ell / \mu$

The Reynolds number is a dimensionless quantity representing the relative importance of inertia forces compared to viscous forces. For large Reynolds numbers the flow is dominated by inertia forces and the fluid motion will most likely be turbulent. At low Reynolds numbers, the viscous forces tend to damp out disturbances, and we can have laminar flow, where the fluid particles slide over each other in smooth filaments or laminae. At intermediate Reynolds numbers, a transition occurs and the laminar changes to turbulent flow. The condition of the surface will determine, to a certain extent, what the transition Reynolds number will be. For instance, a smooth surface can support the laminar development of the boundary layer on a wing for a greater distance than a rough or dirty surface. The advantages of maintaining laminar flow is one of lower drag, since the drag coefficient may be half as much for laminar flow as for turbulent flow. (See Appendix E for a discussion of the drag of laminar and turbulent boundary layers.)

If ρV^2 is proportional to the inertia forces (pressure) and $\mu V/\ell$ is proportional to the viscous forces (shear stress), then the Reynolds number is the ratio

$$Re_\ell = \frac{(\rho V^2)}{(\mu \frac{V}{\ell})} = \frac{\text{inertia forces}}{\text{viscous forces}} = \frac{\rho V \ell}{\mu} .$$

Viscosity, μ, is a function of absolute temperature only,

$$\mu = 2.270 \frac{T^{3/2}}{T + 198.6} \times 10^{-8} \ \frac{\text{slug}}{\text{ft-sec}} ,$$

where T is given in degrees Rankine (degrees Fahrenheit plus 460). The density, ρ, is a function of the pressure and temperature and can be found from the equation of state (see Appendix J).

At standard sea level conditions

$$\mu_{SL} = 3.737 \times 10^{-7} \ \frac{\text{slug}}{\text{ft-sec}}$$

and

$$\rho_{SL} = 0.002377 \ \text{slug/ft}^3 .$$

If we express the airspeed, V, in mph and the characteristic length, ℓ, in feet, the Reynolds number is

$$Re_{\ell_{SL}} = 9324 \ V \text{(mph)} \ \ell \text{(ft)} .$$

If the characteristic length is given in inches, the Reynolds number is

$$Re_{\ell_{SL}} = 777 \ V \text{(mph)} \ \ell \text{(in)} .$$

These expressions for the Reynolds number at sea level based on the length are depicted on the nomogram in Figure I.1.

To find the Reynolds number for an airfoil under stall conditions, we locate the stall speed on the V-scale and the length of the wing chord on the ℓ-scale. Then, connecting the points with a straight line we read the Reynolds number on the middle scale. For 67 mph and 4 foot chord, the Reynolds number at stall is about 2.5 x 10^6 (or 2.5 million). We can find the Reynolds number at other speeds by keeping the chord length fixed and changing the value for V. At a cruise of 180 mph, we find that the Reynolds number is about 6.5 million. These values allow us to choose which sets of experimental data curves to use in selecting the airfoil.

For comparison, the Reynolds number for the Wright Brothers airplane is included. Also plotted for reference is the transition Reynolds number for flow over a flat plate ($Re_{tr} \approx 530{,}000$) indicating the upper limit for laminar boundary layer flow. This can be used to find the extent of laminar flow on the leading part of the wing. (At a given value for airspeed, what is the length ℓ that gives the transition Reynolds number?) The critical Reynolds number for cylinder drag is also included, where the characteristic length is taken to be the cylinder diameter.

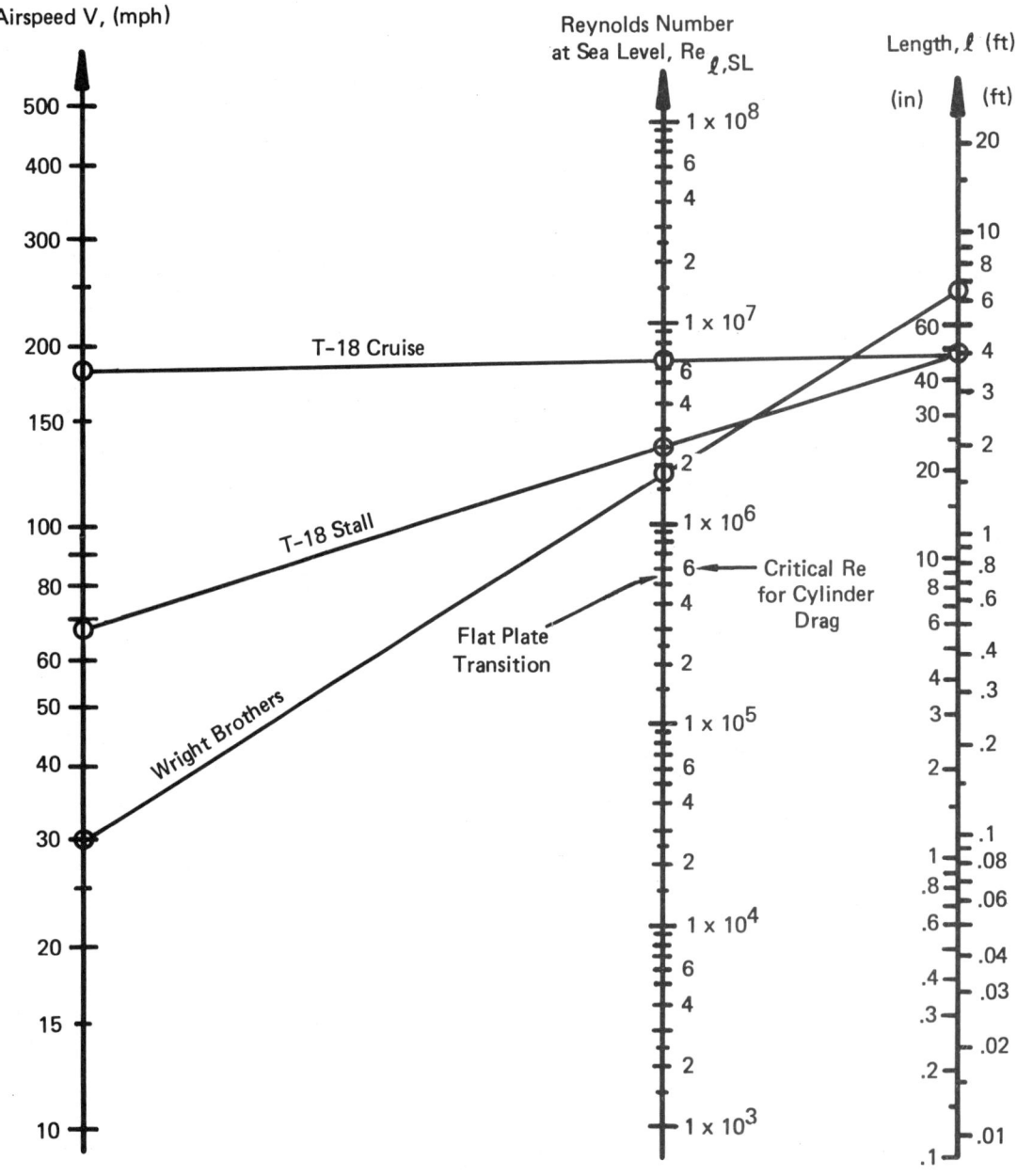

Figure I.1. Nomogram for Calculation of Sea-Level Reynolds Number.

APPENDIX J. EQUATION OF STATE, $p = \rho R T$

The equation of state relates the pressure of a gas to its density and absolute temperature. In the case of air the gas constant, R, is equal to 1718 ft^2/sec^2 °R. The pressure, temperature and density are related to each through the nomogram given in Figure J.1, where the pressure is also given in units of atmospheres and the density also given in terms of the density ratio, $\sigma = \rho/\rho_{SL}$. The density altitude can also be related to the corresponding density ratio (Appendix D). Therefore, we can quickly find the density altitude if we know the pressure and the temperature.

At sea level,

$$p_{SL} = 2116.2 \text{ lb/ft}^2$$

$$\rho_{SL} = 0.002377 \text{ slugs/ft}^3$$

$$T_{SL} = 58.7°F = 518.7°R$$

where the temperature in degrees Rankine is obtained from the temperature in degrees Fahrenheit plus 460.

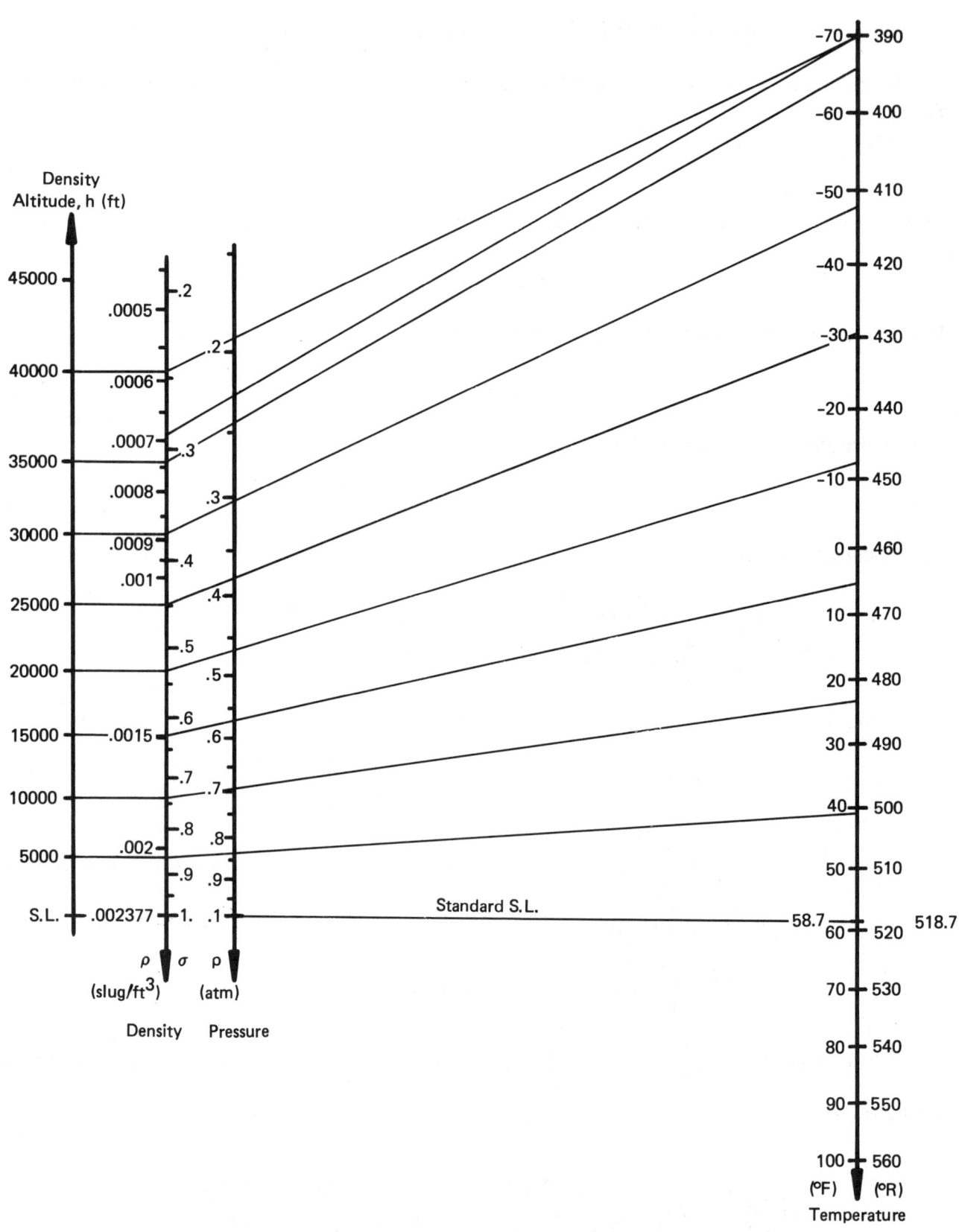

Figure J.1. Nomogram for Equation of State and the Corresponding Density Altitude.

APPENDIX K. HOW TO FIND THE SOLUTION OF A CUBIC EQUATION.

In the course of the idealized propeller analysis, we obtained a cubic equation for the propeller efficiency,

$$\eta^3 + \left(\frac{\pi}{2}\tilde{V}^3\right)\eta - \left(\frac{\pi}{2}\tilde{V}^3\right) = 0 \tag{K.1}$$

where $\tilde{V}$ is a dimensionless airspeed ($= V/V_{prop}$). This is an equation of the form

$$x^3 + ax + b = 0 . \tag{K.2}$$

This form of the cubic equation can also be obtained from the cubic equation given by

$$y^3 + py^2 + qy + r = 0 \tag{K.3}$$

if we substitute $y = x - p/3$, and take

$$a = \frac{1}{3}(3q - p^2) \quad \text{and} \quad b = \frac{1}{27}(2p^3 - 9pq + 27r) . \tag{K.4}$$

To find the solution, let

$$A = \sqrt[3]{-\frac{b}{2} + \sqrt{\frac{b^2}{4} + \frac{a^3}{27}}} \quad \text{and} \quad B = \sqrt[3]{-\frac{b}{2} - \sqrt{\frac{b^2}{4} + \frac{a^3}{27}}} . \tag{K.5}$$

Then, the values of x will be given by

$$x = A + B, \quad -\frac{A+B}{2} + \frac{A-B}{2}\sqrt{-3}, \quad -\frac{A+B}{2} - \frac{A-B}{2}\sqrt{-3} . \tag{K.6}$$

If $\frac{b^2}{4} + \frac{a^3}{27} > 0$,   one real root and two conjugate imaginary roots will exist.

If $\frac{b^2}{4} + \frac{a^3}{27} = 0$,   three roots of which at least two are equal will exist.

If $\frac{b^2}{4} + \frac{a^3}{27} < 0$,   there will be three real and unequal roots.

For our case, $\frac{b^2}{4} + \frac{a^3}{27} = \frac{\left(\frac{\pi}{2}\tilde{V}^3\right)^2}{4}\left[1 + \frac{4}{27}\left(\frac{\pi}{2}\tilde{V}^3\right)\right] > 0$. Therefore, the one root that we are interested in is given by

$$\eta = \left(\frac{\pi}{4}\right)^{1/3}\tilde{V}\left\{\sqrt[3]{1 + \sqrt{1 + \frac{2\pi}{27}\tilde{V}^3}} - \sqrt[3]{-1 + \sqrt{1 + \frac{2\pi}{27}\tilde{V}^3}}\right\} \tag{K.7}$$

REFERENCES AND FURTHER READING (Continued from p. 204)

Anon. 1979), "Advanced Technology Airfoil Research", NASA CP-2045 (2 Parts) and 2046.

Betz, A. (1935), "Applied Airfoil Theory", in *Aerodynamic Theory* edited by W. F. Durand, Volume IV, Division J.

Bingelis, T. (1979), *The Sportplane Builder,* Published by the Author, 8509 Greenflint Lane, Austin, TX 78759.

Dommasch, D. O., S. S. Sherby and T. F. Connolly (1957), *Airplane Aerodynamics,* 2nd Edition, Pitman Publishing Co., NY.

Glauert, H. (1935), "Airplane Propellers", in *Aerodynamic Theory* edited by W. F. Durand, Volume IV, Division L.

Hanson, P. W. (Compiler) (1979), "Science and Technology of Low Speed and Motorless Flight", NASA CP-2085 (Pt I and II).

Hoerner, S. F., and H. V. Borst (1975), *Fluid-Dynamic Lift,* Hoerner Fluid Dynamics, P. O. Box 342, Brick Town, NJ 08723.

Kohlman, D. L. (1979), "Flight Test Results for an Advanced Technology Light Airplane", *J. Aircraft,* Vol. 16, No. 4, pp. 250-255.

Küchemann, D. (1978), *The Aerodynamic Design of Aircraft,* Pergamon Press, Oxford.

Laitone, E. V. (1978a), "Ideal Tail Load for Minimum Aircraft Drag", *J. Aircraft,* Vol. 15, No. 3, pp. 190-190-192.
    (1978b), "Positive Tail Loads for Minimum Induced Drag of Subsonic Aircraft", *J. Aircraft,* Vol. 15, No. 12, pp. 837-842.
    (1979), "Extension of Prandtl's Biplane Theory to Wing-Tail Combinations", *Canadian Aeronautics and Space Journal,* Vol. 25, No. 3, pp. 278-285.
    (1980), "Prandtl's Biplane Theory Applied to Canard and Tandem Aircraft", *J. Aircraft,* Vol. 17, No. 4, pp. 233-237.

Lambie, J. (1980), *Building and Flying Sailplanes and Gliders,* Tab Books, Blue Ridge Summit, PA 17214.

McCormick, B. W. (1967), *Aerodynamics of V/STOL Flight,* Academic Press, NY.

McCormick, B. W. (1979), *Aerodynamics, Aeronautics, and Flight Mechanics,* John Wiley and Sons, NY.

Millikan, C. B. (1941), *Aerodynamics of the Airplane,* John Wiley and Sons, NY.

Milne-Thomson, L. M. (1966), *Theoretical Aerodynamics,* 4th Edition, Macmillian and Co., London (Dover Paperback, 1973).

Von Karman, Th., and J. M. Burgers (1935), "General Aerodynamic Theory — Perfect Fluids", in *Aerodynamic Theory,* edited by W. F. Durand, Volume II, Division E.

Wood, K. D. (1935), *Technical Aerodynamics,* McGraw-Hill, NY.

APPENDIX L. TABULATED PERFORMANCE DATA FOR VARIOUS AIRCRAFT

The performance data for various aircraft have been tabulated in this appendix along with calculated values for the performance rating parameter, F_p, the kinetic energy parameter, WV_{max}^2, and the approximate drag area, multiplied by density ratio and divided by propeller efficiency and power-altitude factor, $\sigma A_D/\eta\phi$. The performance rating parameter is calculated from the expression

$$F_p = \frac{W_u R_{C,max}}{33,000 \, BHP} \left[1 - \frac{V_{min}}{V_{max}} \right] < 1$$

where W_u is the useful load (lb) = $W - W_e$; $R_{C,max}$ is the maximum rate of climb, (fpm); BHP is the engine brake horsepower; V_{min} is the minimum level flight speed (mph); V_{max} is the maximum level flight speed (mph); W is the maximum take off weight (lb); and W_e is the empty weight (lb). The minimum speed was taken to be the stall speed with flaps extended in most cases (except for helicopters and airplanes that are limited by induced drag instead of stall). The data was obtained from **Jane's All The Worlds Aircraft 1977-78** and other sources, including the **1979 Aircraft Directory** by the Editors of **Plane and Pilot**, some aircraft manuals, and various sets of theoretical calculations.

No changes were made in the numbers supplied by the manufacturers. Therefore, direct comparisons of the performance rating parameter, F_p, should be made with care.

In addition to the performance rating parameter, the kinetic energy parameter, WV_{max}^2, is calculated for the various aircraft. F_p vs WV_{max}^2 is plotted for each of the aircraft in Figure 67 in Part 1 of the text. If your airplane does not appear on the list, the end of the table has room to fill in the appropriate data and calculate the numbers in the last three columns. Then the data point can be plotted and compared to the other airplanes of this section.

The third quantity of interest is the estimate of the drag area from the sea-level engine brake horsepower and the maximum level airspeed. The modified drag area. $\sigma A_D/\eta\phi$ (with A_D in ft^2), is calculated from the equation

$$\frac{\sigma A_D}{\eta\phi} = \frac{146625 \, BHP_{SL}}{V_{max}^3}$$

where the density ratio, σ, and the power-altitude factor, ϕ, are, unity for sea level and can be found from Appendix D or Figure 43 for other altitudes. This equation is only an approximation because the contribution of the induced drag has been neglected. If the maximum speed is much larger than the speed for minimum sink rate (or stall speed if that is the limiting airspeed) the maximum airspeed will be proportional to the cube-root of the available horsepower and the expression will be more accurate. If the airplane is underpowered, it will run out of power before the induced drag becomes negligible and the calculated drag area will be smaller than is calculated.

The data for $\sigma A_D/\eta\phi$ is plotted in Figure 12 of Part 1 and is used as follows: (1) Locate the airplane and find the value of $\sigma A_D/\eta\phi$; (2) estimate the propeller efficiency (η = 0.8 is a good first guess) and the ratio ϕ/σ from Figure 43 for the desired altitude; (3) multiply each of these values times $\sigma A_D/\eta\phi$ to find an approximation for the drag area, A_D; (4) if it is reasoned that the induced drag is still important for the speed to which the data applies, take some fraction of this drag area for the final value of A_D. A feel for the magnitude of the drag area will be helpful when it comes time to calculate the performance for the next design problem.

Other quantities that may be of use for a sample calculation for the performance of the airplane using the Airplane Performance and Design Nomogram are: b, the wing span (in feet); D_p, the propeller diameter (in feet and inches); and S, the wing area (in square feet).

Even though the performance data for each airplane are not complete, the table was left as is so that the remaining quantities may be filled in as they become available. The list is somewhat subjective; all of the possible airplanes of the world are not given. That is one reason for the additional empty space at the beginning and end of this table — you can put in your missing favorites.

	BHP (hp)	b (ft)	D_p (in)	S (ft²)	W_e (lb)	W (lb)	V_{max} (mph)	$V_{s,0}$ (mph)	$R_{C,max}$ (fpm)	W_u (lb)	WV^2_{max} (lb mph²)	F_p (—)	$\dfrac{\sigma A_D}{\eta \phi}$ (ft²)

	FACTORY BUILT AIRPLANES	BHP (hp)	b (ft)	D_p (in)	S (ft^2)	W_e (lb)	W (lb)	V_{max} (mph)	$V_{s,0}$ (mph)	$R_{C,max}$ (fpm)	W_u (lb)	WV^2_{max} (lb mph^2)	F_p (—)	$\frac{\sigma A_D}{\eta \phi}$ (ft^2)
1.	Beechcraft Sierra 200	200	32'9"	6'4"	146	1693	2750	163	69	928	1057	7.3 × 10^7	.086	6.8
2.	Beechcraft Sundowner 180	180	32'9"	6'4"	146	1477	2450	141	59	792	973	4.8 × 10^7	.075	9.4
3.	Beechcraft Sport 150	150	32'9"	6'2"	146	1414	2150	127	58	680	736	3.5 × 10^7	.055	10.7
4.	Beechcraft Bonanza V35B	285	33'6"	7'0"	181	2051	3400	209	59	1167	1349	1.5 × 10^8	.120	4.6
5.	Beechcraft Bonanza F33A	285	33'6"	7'0"	181	2112	3400	209	59	1167	1288	1.5 × 10^8	.115	4.6
6.	Beechcraft Bonanza A36	285	33'6"	7'0"	181	2157	3600	206	60	1030	1443	1.5 × 10^8	.112	4.8
7.	Beechcraft Baron 95-B55	520	37'10"	6'6"	199.2	3226	5100	231	84	1693	1874	2.7 × 10^8	.118	6.2
8.	Beechcraft Duke B60	760	39'3"	6'2"	212.9	4380	6775	283*	84	1601	2395	5.4 × 10^8	.108	4.9
9.	Beechcraft Queen Air B80	760	50'3"	7'9"	293.9	5277	8800	248†	81	1275	3523	5.4 × 10^8	.121	7.3
10.	Beechcraft King Air C90	1100	50'3"	7'9"	293.94	5717	9650	256#	87	1955	3933	6.3 × 10^8	.140	9.6
11.	Bellanca 17-30A Super Vik 300A	300	34'2"	6'8"	161.5	2217	3325	188¶	70	1085	1108	1.2 × 10^8	.076	6.6
12.	Bellanca Citabria 7ECA	115	33'5"		165	1067	1650	117	51	725	583	2.3 × 10^7	.063	10.5
13.	Bellanca 8KCAB Decathlon	150	32'0"		170	1280	1800	135●	53	880	520	3.3 × 10^7	.056	8.9
14.	Bellanca 19-25 Skyrocket II	435	35'0"	6'10"	182.6	2300	4100	331§	65	1900	1800	4.5 × 10^8	.191	1.76
15.	Cessna 150	100	32'8.5"	5'9"	157	1000	1600	125	48	670	600	2.5 × 10^7	.075	7.5
16.	Cessna Skyhawk/100	160	35'10"	6'3"	174	1379	2300	144	51	770	921	4.8 × 10^7	.087	7.9
17.	Cessna Cardinal	180	35'6"	6'4"	174	1533	2500	160	53	840	967	6.4 × 10^7	.091	6.4
18.	Cessna Cardinal RG	200	35'6"	6'6"	174	1703	2800	180	57	925	1097	9.1 × 10^7	.105	5.0
19.	Cessna 180 Skywagon	230	35'10"	6'10"	174	1648	2800	170	55	1100	1152	8.1 × 10^7	.113	6.9
20.	Cessna Skylane	230	35'10"	6'10"	174	1717	2950	170	57	1010	1233	8.5 × 10^7	.109	6.9
21.	Cessna 185 Skywagon	300	35'10"	6'10"	174	1687	3350	178	56	1010	1663	1.1 × 10^8	.116	7.8
22.	Cessna Stationair	300	35'10"	6'8"	174	1808	3600	180	62.5	920	1792	1.2 × 10^8	.109	7.5
23.	Cessna Skywagon 207	300	35'10"	6'8"	174	1996	3800	173	67	810	1804	1.1 × 10^8	.090	8.5
24.	Cessna 210 Centurion	300	36'9"	6'8"	175	2175	3800	202	64.5	860	1625	1.6 × 10^8	.096	5.3
25.	Cessna 310	570	36'11"	6' 4.5"	179	3347	5500	238	81	1662	2153	3.1 × 10^8	.125	6.2

* — 23,000 ft.   † — 11,500 ft.   # — 12,000 ft.   ¶ — Cruising Speed   ● — 75% Power at 8,000 ft.   § — 29,000 ft

	FACTORY BUILT AIRPLANES	BHP (hp)	b (ft)	D_p (in)	S (ft^2)	W_e (lb)	W (lb)	V_{max} (mph)	$V_{s,0}$ (mph)	$R_{C,max}$ (fpm)	W_u (lb)	WV_{max}^2 (lb mph^2)	F_p (—)	$\frac{\sigma A_D}{\eta \phi}$ (ft^2)
26.	Cessna 337 Skymaster	420	38'2"	6'6" 6'4"	202.5	2790	4630	206	70	1100	1840	2.0 x 10^8	.096	7.0
27.	Cessna 340A	620	38' 1.3"	6' 4.5"	184	3878	5990	281*	82	1650	2112	4.7 x 10^8	.121	4.1
28.	Cessna 402 Utiliner	600	39' 10.25"	6' 4.5"	195.72	3904	6300	264†	81	1610	2396	4.4 x 10^8	.135	4.8
29.	Cessna 414 Chancellor	620	39'11"	6' 4.5"	195.72	4135	6350	275*	81	1580	2215	4.8 x 10^8	.121	4.4
30.	Cessna Titan Ambassador	750	46'4"	7'6"	242	4804	8400	267†	81	1575	3596	6.0 x 10^8	.159	5.8
31.	Great Lakes 2T-1A-1	140	26'8"		187.6	1140	1750	120	54	1000	610	2.5 x 10^7	.073	11.9
32.	Great Lakes 2T-1A-2	180	26'8"		187.6	1230	1800	132	57	1400	570	3.1 x 10^7	.076	11.5
33.	Grumman American Lynx	115	24' 5.5"	6'0"	100.92	1066	1600	136	60	750	534	3.0 x 10^7	.059	6.7
34.	Grumman American Cheetah	150	31'6"	6'1"	140.12	1303	2200	157	61	660	897	5.4 x 10^7	.073	5.7
35.	Grumman American Tiger	180	31'6"	6'1"	140.12	1311	2400	170	61	850	1084	6.9 x 10^7	.100	5.4
36.	Grumman American GA-7 Cougar	320	36' 10.25"	6'1"	184		3800	200	69	1200		1.5 x 10^8		5.9
37.	Helio Super Courier H-295	295	39'0"	8'0"	231	2080	3400	167	30	1150	1320	9.5 x 10^7	.128	9.3
38.	Lake LA-4-200 Buckaneer	200	38'0"	6'2"	170	1550	2960	146	45	1200	1140	5.7 x 10^7	.143	9.4
39.	Maule M-5-210C Lunar Rocket	210	30'10"	6'2"	157.9	1325	2450	158	56	1250	1125	6.1 x 10^7	.131	7.8
40.	Maule M-5-235C Lunar Rocket	235	30'10"	6'6"	157.9	1375	2450	172	56	1350	1075	7.2 x 10^7	.126	6.8
41.	Meyers 200D	285	30'6"	6'10"	161.5	1940	3400	215	54	1400	1460	1.6 x 10^8	.163	4.2
42.	Mooney Ranger	180	35'0"	6'2"	167	1525	2575	169	56	800	1050	7.4 x 10^7	.095	5.5
43.	Mooney Executive	200	35'0"	6'2"	167	1640	2740	177	62	1005	1100	8.6 x 10^7	.109	5.3
44.	Mooney 201	200	35'0"	6'2"	167	1640	2740	202	63.5	1030	1100	1.1 x 10^8	.118	3.6
45.	Piper PA-18 Super Cub 150	150	35'2½"		178.5	946	1750	130#	43	960	804	3.0 x 10^7	.104	10.0
46.	Piper PA-23-250 Aztec F	500	37'2½"	6'5"	207.6	3221	5200	215	68	1400	1979	2.4 x 10^8	.115	7.4
47.	Piper Cherokee Cruiser	150	35'0"	6'2"	170	1290	2150	143	47	631	860	4.4 x 10^7	.074	7.5
48.	Piper PA-28-161 Cherokee Warrior II	160	35'0"	6'2"	170	1336	2325	140	57.5	710	989	4.6 x 10^7	.078	8.5
49.	Piper PA-28-181 Cherokee Archer II	180	35'0"	6'4"	170	1416	2550	147	56.5	740	1134	5.5 x 10^7	.087	8.3
50.	Piper PA-28R1-201 Cherokee Arrow III	200	35'0"	6'2"	170	1622	2750	175	63.5	831	1128	8.4 x 10^7	.090	5.5

* — 20,000 ft.   † — 16,000 ft.   # — 75% Power

	FACTORY BUILT AIRPLANES	BHP (hp)	b (ft)	D_p (in)	S (ft^2)	W_e (lb)	W (lb)	V_{max} (mph)	$V_{s,0}$ (mph)	$R_{C,max}$ (fpm)	W_u (lb)	WV^2_{max} (lb mph^2)	F_p (—)	$\dfrac{\sigma A_D}{\eta \phi}$ (ft^2)
51.	Piper PA-28-235 Pathfinder	235	32'0"		168	1592	3000	161	61	800	1408	7.8 x 10^7	.090	8.3
52.	Piper PA-32 Cherokee Six	300	32'8¾"	6'8"	174.5	1784	3400	174	54	1050	1616	9.9 x 10^7	.118	8.3
53.	Piper PA-32R-300 Cherokee Lance	300	32' 9⅞"	6'8"	174.5	1973	3600	190	60	1000	1627	1.3 x 10^8	.112	6.4
54.	Piper PA-31-310 Navajo C.	620	40'8"	6'8"	229	3991	6500	261*	73	1445	2509	4.4 x 10^8	.128	5.1
55.	Piper PA-31-350 Navajo Chieftain	700	40'8"	6'8"	229	4219	7000	270*	85	1390	2781	5.1 x 10^8	.115	5.2
56.	Piper PA-31T Cheyenne	1240	42'8¼"	7'9"	229	4976	9000	326†	88	2800	4024	9.6 x 10^8	.201	5.2
57.	Piper PA-34 Seneca II	400	38'10¾"	6'4"	208.7	2823	4570	225#	70.5	1340	1747	2.3 x 10^8	.122	5.1
58.	Pitts S-1	180	17'4"	6'4"	98.5	720	1150	176	62	2600	430	3.6 x 10^7	.122	4.8
59.	Pitts S-2A	200	20'0" / 19'0"		125	1000	1575	157	59	1800	575	3.9 x 10^7	.098	7.6
60.	Rockwell Bronco OV-10A	1430	40'0"		291	6969	9908	281	74	2665	2939	7.8 x 10^8	.122	9.4
61.	Rockwell Commander 112B	200	35' 7.2"	6'5"	163.8	1773	2800	173	61	880	1027	8.4 x 10^7	.089	5.7
62.	Rockwell Commander 114	260	32'9"	6'5"	152	1858	3140	189	62.5	1088	1282	1.1 x 10^8	.109	5.6
63.	Rockwell Shrike Commander 500S	580	49'.5"	6'8"	255	4635	6750	215	68	1340	2115	3.1 x 10^8	.101	8.6
64.	Rockwell Turbo Commander 690B	1400	46'8"	8'10"	266	6195	10,325	330#	89	2821	4130	1.1 x 10^9	.184	5.7
65.	Rockwell Commander 700	650	42' 5.5"	6.9"	200.2	4400	6600	266¶	80	1460	2200	4.7 x 10^8	.105	5.1
66.	Swearingen Merlin IIIA	1680	46'3"	8'2"	277.5	7400	12,500	325●	96	2530	5100	1.3 x 10^9	.164	7.2
67.	Swearingen Merlin IVA	1880	46'3"	8'6"	277.5	8200	12,500	310●	99	2400	4300	1.2 x 10^9	.113	9.3
68.	Swearingen SA-226TC-Metro II	1880	46'3"	8'6"	277.5	7450	12,500	294§	99	2400	5050	1.1 x 10^9	.130	10.8
69.	Taylorcraft F-19 Sportsman 100	100	36'0"		183.71	900	1500	127	43	775	600	2.4 x 10^7	.093	7.2
70.	Ted Smith Aerostar 600	580	34'2"	6'6"	170	3720	5500	260	77	1800	1780	3.7 x 10^8	.118	4.8
71.	Univair Stinson 108-5	180	33'11"	6'4"	155	1300	2400	152	61	1000	1100	5.5 x 10^7	.111	7.5
72.	Varga 2150A Kachina	150	30'0"	6'2"	142	1125	1817	148	52	1450	692	4.0 x 10^7	.131	6.8
73.	Lockheed Model 82 Hercules	18032	132'7"	13'6"	1745	75331	155,000	386	115	1900	79669	2.3 x 10^{10}	.179	46.0
74.	Lockheed Model 85 Orion	19640	99'8"	13'6"	1300	61491	135,000	476	129	1950	73509	3.1 x 10^{10}	.161	26.7

* — 15,000 ft.   † — 11,000 ft.   # — 12,000 ft.   ¶ — 20,000 ft.   ● — 16,000 ft.   § — 10,000 ft.

		BHP (hp)	b (ft)	D_p (in)	S (ft^2)	W_e (lb)	W (lb)	V_{max} (mph)	$V_{s,0}$ (mph)	$R_{C,max}$ (fpm)	W_u (lb)	WV^2_{max} (lb mph^2)	F_p (—)	$\dfrac{\sigma A_D}{\eta \phi}$ (ft^2)

	MILITARY AIRCRAFT	BHP (hp)	b (ft)	D_p (in)	S (ft^2)	W_e (lb)	W (lb)	V_{max} (mph)	$V_{s,0}$ (mph)	$R_{C,max}$ (fpm)	W_u (lb)	WV_{max}^2 (lb mph^2)	F_p (—)	$\frac{\sigma A_D}{\eta \phi}$ (ft^2)
1.	Bell P-63 Kingcobra	1325				6375	8800	410		3500	2425	1.5 x 10^9		2.8
2.	Chance Vought F4U Corsair	2250				8694	12039	425	80	3120	3345	2.2 x 10^9	.114	4.3
3.	Curtis-Wright P-40 Hawk	1200				5381	6789	365	72		1408	9.0 x 10^8		3.6
4.	Douglas Dauntless	1200				6450	10500	255	78	1428	4050	6.8 x 10^8	.101	10.6
5.	Grumman F4F Wildcat	1000				4425	5876	325	66	3300	1451	6.2 x 10^8	.116	4.3
6.	Grumman Hellcat	2000				9153	12500	400		3000	3347	2 x 10^9		4.6
7.	Grumman F8F Bearcat	2100				7070	9300	425		5700	2230	1.7 x 10^9		4.0
8.	Hawker Sea Fury	2560				9240	12350	450		4320	3110	2.5 x 10^9		4.1
9.	Messerschmitt BF-109	1200				4180	6090	390		3320	1910	9.3 x 10^8		3.0
10.	North American P-51 Mustang	1720				7125	11600	437	80	3475	4475	2.2 x 10^9	.224	3.0
11.	North American AT-6 Texan	550				4158	5300	205	67	1600	1142	2.2 x 10^8	.068	9.4
12.	Republic P-47 Thunderbolt	2300				9900	12500	440	100	2500	2600	2.4 x 10^9	.066	4.0
13.														
14.														
15.														
16.														
17.														
18.														
19.														
20.														
21.														
22.														
23.														
24.														

		BHP (hp)	b (ft)	D_p (in)	S (ft²)	W_e (lb)	W (lb)	V_{max} (mph)	$V_{s,0}$ (mph)	$R_{C,max}$ (fpm)	W_u (lb)	WV_{max}^2 (lb mph²)	F_p (—)	$\frac{\sigma A_D}{\eta \phi}$ (ft²)

	HOME BUILT AIRPLANES	BHP (hp)	b (ft)	D_p (in)	S (ft²)	W_e (lb)	W (lb)	V_{max} (mph)	$V_{s,0}$ (mph)	$R_{C,max}$ (fpm)	W_u (lb)	WV_{max}^2 (lb mph²)	F_p (—)	$\frac{\sigma A_D}{\eta \phi}$ (ft²)
1.	Aerocar Imp		29'0''	6'0''	112	950	1550	150	50	800	600	3.5×10^7		
2.	Aerocar Mini Imp	68	25'6''	4'9''	76.5	520	800	150	45	1200	280	1.8×10^7	.105	3.0
3.	Aeroneering Miller Lil' Rascal	85	20'8''	5'9''	126	650	1100	100	55		450	1.1×10^7		12.5
4.	Aerosport Woody Pusher	75	29'0''		130	630	1150	98	45	600	520	1.1×10^7	.068	11.7
5.	Aerosport Rail Mk II	48	23' 3.5''	4'6''	81.5	446	730	90	45	1100	284	5.9×10^6	.099	9.7
6.	Am. Air Racing JP001 Wild Turkey	100	20' 3.75''	4'10''	67	510	750	270	70	1850	240	5.5×10^7	.100	.74
7.	Anderson EA-1 Kingfisher	100	36'1''	6'0''		1032	1500	120	45	550	468	2.2×10^7	.049	8.5
8.	Bakeng Duce	125	30'4''	6'2''	138	898	1500	145	36	2000	602	3.2×10^7	.214	6.0
9.	Barnett J-3M	65	--		--	400	650	85	0	500	250	4.7×10^6	.058	15.5
10.	Barney Oldfield 'Baby' Lakes	80	16'8''		86	475	850	135	50	2000	375	1.5×10^7	.179	4.8
11.	Beaujon Flybike	8	28' 7.75''		92	145	375	50	26	200	230	9.4×10^5	.084	9.4
12.	Bede BD-4	180	25'7''	6'4''	102.33	1080	2000	183	61	1400	920	6.7×10^7	.145	4.3
13.	Bede BD-5 Micro	70	17'0''		38	410	850	230	69	1750	440	4.5×10^7	.233	.84
14.	Bede BD-6	55	21'6''		55.5	375	650	140	50	900	275	1.3×10^7	.088	2.9
15.	Bede BD-7	140	24'0''	5'0''	93'5	960	2000	220	65	1500	1040	9.7×10^7	.238	1.9
16.	Bede BD-8	200	19'4''	6'2''	96.67	820	1300	210	48	2200	480	5.7×10^7	.123	3.2
17.	Beets G/B Special	70	25'0''		104.2	603	925	156	35	2000	322	2.3×10^7	.216	2.7
18.	Benson B-8M	72	--	4'0''	--	247	500	85	15	1000	253	3.6×10^6	.088	17.2
19.	Birdman TL-1	15	34'0''	2'4''	144.5	100	350	48	18	200	250	8.1×10^5	.063	19.9
20.	Bushby/Long MM-1-85 Midget Mustang	85	18'6''		68	575	875	190	57	1750	300	3.2×10^7	.131	1.8
21.	Bushby/Long MM-1-125 Midget Mustang	135	18'6''		68	590	900	225	60	2200	310	4.6×10^7	.112	1.7
22.	Cassutt Special I	85	14'11''		66	516	730	230	70	2000	214	3.9×10^7	.106	1.0
23.	Cassutt Special II	85	13'8''		66	433	800	235	62	3000	367	4.4×10^7	.289	.96
24.	Condor Shoestring	100	19'0''		66	565	800	240	65	3000	235	4.6×10^7	.156	1.1

	HOME BUILT AIRPLANES	BHP (hp)	b (ft)	D_p/BHP	S (ft²)	W_e (lb)	W (lb)	V_{max} (mph)	$V_{s,0}$ (mph)	$R_{C,max}$ (fpm)	W_u (lb)	WV^2_{max} (lb mph²)	F_p (—)	$\sigma A_D / \eta \varphi$ (ft²)
25.	Cvjetkovic CA-61	65	27'6"		126.5	800	1300	120	50		500	1.9 × 10⁷		5.5
26.	Cvjetkovic CA-65	125	25'0"	5'8"	108	900	1500	160	55	1000	600	3.8 × 10⁷	.095	4.5
27.	Cvjetkovic CA-65A	150	25'5"	5'8"	109.4	900	1500	174	55	1530	600	4.5 × 10⁷	.127	4.2
28.	d'Apuzzo D-260(3) Senior Sport	260	27'0"		185		2150	155*	55	2000		5.2 × 10⁷		10.2
29.	d'Apuzzo D-200 Junior Aero Sport	180	21'8"		140	840	1275	160*	55	2500	430	3.3 × 10⁷	.119	6.4
30.	Davis DA-2A	65	19' 2.75"		82.5	610	1125	120			515	1.6 × 10⁷		5.5
31.	Davis DA-5A	65	15' 7.25"	5'0"	57.2	460	775	160	60	800	315	2.0 × 10⁷	.073	2.3
32.	DSK Airmotive DSK-1 Hawk	65	20' 4.5"	5'3"	64	525	893	146†	50	1500	368	1.9 × 10⁷	.169	3.1
33.	DSK Airmotive DSK-2 Golden Hawk	65	20' 4.5"	5'3"	64.5	550	914	156†	50	1500	364	2.2 × 10⁷	.173	2.5
34.	Dyke JD-2 Delta	180	22' 2.5"	6'2"	183	960	1900	190#		2000	940	6.9 × 10⁷		3.8
35.	EAA Biplane	85	20'0"	5'10"	108	710	1150	125	55	1000	440	1.8 × 10⁷	.088	6.4
36.	EAA Acro-Sport	180	19'7"	6'4"	115.5	733	1178	152	50	3500	445	2.7 × 10⁷	.176	7.5
37.	EAA Super Acro-Sport	200	19'7"	6'4"	115.5	884	1350	156	50	3700	466	3.3 × 10⁷	.178	7.7
38.	EAA Pober Pixie	60	29'10"	4'5"	134.75	526	900	103	30	700	373	9.5 × 10⁶	.093	8.1
39.	EAA Nesmith Cougar	115	20'6"		82.5	624	1250	195	53	1300	626	4.8 × 10⁷	.156	2.3
40.	Evans VP-1	60	24'0"	4'6"	100	440	750		40	400	310			
41.	Evans VP-2	60	27'0"	5'0"	130	640	1040	100	40	400	400	1.0 × 10⁷	.048	8.8
42.	Fike "E"	85	22' 4½"	6'2"	143.1	690	1100	120	35	1000	410	1.6 × 10⁷	.104	7.2
43.	Flaglar Slooter	40	28'0"	4'6"	115	390	650	90	34	600	260	5.3 × 10⁶	.074	8.0
44.	Flight Dynamics Flightsail VII	90	39'0"	6'2"	195	1200	1700	95	40	500	500	1.5 × 10⁷	.048	15.4
45.	Forbes F-3 Cobra	170	19'6"	5'8"	77.7	675	1100	207	58	3000	425	4.7 × 10⁷	.164	2.8
46.	Franks You-Two J12		34'0"			440	680	60		400	240	2.4 × 10⁶		
47.	Grega GN-1 Aircamper	65	29'0"		150		1129	95	35	600		1.0 × 10⁷		11.1
48.	Harmon der Domerschlag	75	19'6"	4'2"	78	350	600	120	55	800	250	8.6 × 10⁶	.044	6.4

* − 7,000 ft. † − 5,000 ft. # − 7,500 ft.

	HOME BUILT AIRPLANES	BHP (hp)	b (ft)	D_p (in)	S (ft²)	W_e (lb)	W (lb)	V_{max} (mph)	$V_{s,0}$ (mph)	$R_{C,max}$ (fpm)	W_u (lb)	WV_{max}^2 (lb mph²)	F_p (—)	$\frac{\sigma A_D}{\eta \phi}$ (ft²)
49.	Harmon 1-2 Mister America	65	19'8"	4'6"	76	430	650	125	48	800	230	1.0×10^7	.053	6.4
50.	Hatz CB-1 Biplane	150	26'0"	6'2"	190	966	1600	100*	45	1200	634	1.6×10^7	.085	22.0
51.	Hollmann HA-2M Sportster	130	—	5'6"	—	620	1050	121*	28	500	430	1.5×10^7	.039	10.8
52.	Hovey Wing Ding II	14	17'0"	4'0"	98	123	310	50	26		187	7.8×10^5		16.4
53.	Javelin Wichawk	180	24'0"	6'4"	185	1280	2000	140		1700	720	3.9×10^7		9.6
54.	Jeffair Barracuda	220	24'9"	7'2"	120	1495	2300	218†	62	2200	805	1.1×10^8	.175	3.1
55.	Keleher Lark-1B	65	23'0"	5'7"	80.5	550	855	132	55	900	305	1.5×10^7	.075	4.1
56.	K-1 Kraft Super Fli	200	24'0"			980	1400	200		3000	420	5.6×10^7		3.7
57.	Larkin KC-3 Skylark	100	26'6"	6'0"	114	790	1246	115	42	550	456	1.6×10^7	.048	9.6
58.	McCarley Mini-Mac	55#	20'6"		68.3	514	800	160		1000	286	2.0×10^7		2.0
59.	MacDonald S-20	53	25'0"	4'5"	94	456	720	110	38	850	264	8.7×10^6	.084	5.8
60.	Merkel Mark II	220	25'6"	6'8"	146	1200	1540	163	55	2500	340	4.1×10^7	.078	7.4
61.	Miller JM-2	100	15'0"	3'3"	66	630	1100	235	74	1600	470	6.1×10^7	.156	1.1
62.	Miller WM-2	65	40'0"	6'2"	144	775	1050	136	45	890	275	1.9×10^7	.076	3.8
63.	Mini-Hawk TH.E.01 Tiger Hawk	72	18'0"	4'6"	57	525	800	175	50	1000	275	2.5×10^7	.083	2.0
64.	Monnett Sonerai	60	16'8"	4'2"	75	440	700	175	46		260	2.1×10^7		1.6
65.	Monnett Sonerai II	70	18'8"	4'6"	84	506	925	165	45	750	419	2.5×10^7	.099	2.3
66.	Osprey II	150	26'0"	5'6"	130	970	1560	130¶	60	1000	590	2.6×10^7	.064	7.5
67.	Stevenson/Owl OR-70 Fang	100	20'0"	4'10"	66	580	840	250	40	3000	260	5.3×10^7	.199	.94
68.	Vin-Del/Owl OR-71 Lil Quickie	100	20'0"	5'0"	66	553	850	255	69	3000	297	5.6×10^7	.197	.89
69.	Parker Teenie Two	42	18'0"			310	590	120	50	800	280	8.5×10^6	.094	3.6
70.	Payne Knight Twister KT-85	90	15'0"		60	535	960	160	60	900	425	2.5×10^7	.080	3.2
71.	Pazmany PL-1 Laminar	95	28'0"		116	800	1326	120	51	1000	526	1.9×10^7	.096	8.1
72.	Pazmany PL-2	150	28'0"		116	902	1447	153	54	1700	545	3.4×10^7	.121	6.1

* — Max. Cruising Speed † — 7,000 ft. # — Calculated from 1834 cc ¶ — 75% Power

	HOME BUILT AIRPLANES	BHP (hp)	b (ft)	D_p (in)	S (ft²)	W_e (lb)	W (lb)	V_{max} (mph)	$V_{s,0}$ (mph)	$R_{C,max}$ (fpm)	W_u (lb)	WV_{max}^2 (lb mph²)	F_p (—)	$\frac{\sigma A_D}{\eta \phi}$ (ft²)
73.	Pazmany PL-4A	50	26'8"	5'8"	89	578	850	125	48	650	272	1.3 × 10⁷	.066	3.8
74.	PDQ-2	40	22'0"			360	600	80	40	500	240	3.8 × 10⁶	.045	11.5
75.	Pietenpol B4 Aircamper	60	29'0"	5'0"	145	622		110	38	500				6.6
76.	Powell P-70 Acey Ducey	65	32'6"		155	750	1275	104	30	625	525	1.4 × 10⁷	.109	8.5
77.	R & B Aircraft Grasshopper	65	30'0"				575		25	1000				
78.	Rand Robinson KR-1	58	17'2"	4'5"	64	380	900	200	45	900	520	3.6 × 10⁷	.189	1.1
79.	Rand Robinson KR-2	60	20'8"	4'4"	80	440	800	180	45	800	360	2.6 × 10⁷	.109	1.5
80.	RLU-1 Breezy	90	33'0"		165	700	1200	75 *	30		500	6.8 × 10⁶		21.9
81.	Rotor Way Scorpion Too	133	— —	25'0"	— —	805	1235			800	430			
82.	Rutan Van Viggen	150	19'0"	5'10"	119	950	1700	163		1200	750	4.5 × 10⁷		5.1
83.	Rutan Varieze	100	22'4"		67	535	1050		55.5	1700	515			
84.	Sequoia 300 Side-By-Side	300	30'0"	6'8"	130	1700	3000	230	69	1600	1300	1.6 × 10⁸	.147	3.6
85.	Sequoia 301 Tandem	300	29'2"	6'8"	125	1700	3000	239	69	1600	1300	1.7 × 10⁸	.149	3.2
86.	Sindlinger HH-1 Hawker Hurricane	150	25'0"	6'4"	101	1005	1375	200	62	1850	370	5.5 × 10⁷	.095	2.7
87.	Skyote	118	20'0"	6'2"	123	595	895	115	44	1900	300	1.2 × 10⁷	.090	11.4
88.	Smith DSA-1 Miniplane	108	17'0"	5'11'	100	616	1000	135	55	1250	384	1.8 × 10⁷	.080	6.4
89.	Smyth 'S' Sidewinder	125	24'10"	5'7"	96	867	1450	185	55	900	583	5.0 × 10⁷	.089	2.9
90.	Sorrell Hiperbipe	180	22'10"	6'4"	150	1236	1911	172	49	1500	675	5.7 × 10⁷	.122	5.2
91.	Spencer Amphibian Aircraft S-12-E	285	37'4"	7'0"	184	2190	3200	147	43	1000	1010	6.9 × 10⁷	.076	13.2
92.	Spezio DAL-1 Tuholder	150	24'9"		120.7	900	1500	150	55	2400	600	3.4 × 10⁷	.184	6.5
93.	Sport A/C Minicoupe	65	24'0"	4'6"	835	497	850	105	48	750	353	9.4 × 10⁶	.067	8.2
94.	Stephens Akro	180	24'6"	6'4"	100	950	1300	170	55	4000	350	3.8 × 10⁷	.160	5.4
95.	Stewart Headwings	36	28'3"	5'2"	110.95	437	700	80	38	400	263	4.5 × 10⁶	.046	10.3
96.	Stewart Foo Fighter	125	20'8"	6'0"	130	725	1100	115 †	48	1200	375	1.5 × 10⁷	.064	12.1

* — 70% Power † — Max. Cruising Speed

	HOME BUILT AIRPLANES	BHP (hp)	b (ft)	D_p (in)	S (ft²)	W_e (lb)	W (lb)	V_{max} (mph)	$V_{s,0}$ (mph)	$R_{C,max}$ (fpm)	W_u (lb)	WV^2_{max} (lb mph²)	F_p (—)	$\frac{\sigma A_D}{\eta \phi}$ (ft²)
97.	Stolp SA-300 Standuster Too	180	24'0"		192	1105	1650	180		2600	545	5.3×10^7		4.5
98.	Stolp SA-500 Starlet	100	25'0"		83		750							
99.	Stolp SA-700 Acroduster I	180	19'0"		105	740	1190	180	70	3000	450	3.9×10^7	.139	4.5
100.	Stolp SA-750 Acroduster II	200	21'5"		130				55	2300				
101.	Stolp SA-900 V-Star	65	23'0"		141				35	600				
102.	Taylor Tinker Toy	140	30'4"		136.5	1170	1650	124		1000	480	2.5×10^7		10.8
103.	Turner T-40A	85	25'2"		89.9	82.8	1410	150	54	750	582	3.2×10^7	.100	3.7
104.	Turner Super T-40A	125	26'8"		102.5	980	1550	175	50	1400	570	4.7×10^7	.138	3.4
105.	Turner T-40C	150	28'0"		102	828	1650	190	47	1500	822	6.0×10^7	.187	3.2
106.	Thorp T-18 Tiger	180	20'10"	5'3"	86	900	1506	200	65	2000	606	6.0×10^7	.138	3.3
107.	Tru-Flite TF-1 Tomcat	160	24'0"	6'2"	133	1154	1725	160	55	1100	571	4.4×10^7	.078	5.7
108.	Turner (Carroll) CP-16	85	—	4'9"	—			96						
109.	Van's RV-3	125	19'11"	5'8"	90	695	1050	195	48	1900	355	4.0×10^7	.123	2.5
110.	Volmer VJ-22 Sportsman	85	36'6"		175	1000	1500	95	45	600	500	1.4×10^7	.056	14.5
111.	WAR Focke-Wulf 190	100	20'0"	5'0"	70	630	900	185	55	1000	270	3.1×10^7	.057	2.3
112.	Wendt WH-1 Traveler	75	30'0"	5'11"	118	900	1400	131	57	500	500	7.4×10^7	.057	4.9
113.	Westfall Special W-7	125	23'10"		83.4	1157	1747	140	15	1500	590	3.4×10^7	.192	6.7
114.	White WW-1 der Jäger D.IX	115	20'0"	5'6"	115	534	888	145	54	2400	354	1.9×10^7	.140	5.5
115.	Wittman Tailwind W-8	100	22'6"	5'4"	90	700	1300	165	55	900	600	3.5×10^7	.109	3.3
116.	Zinno Olympian ZB-1	*.48	78'6"	8'8"	312	148	290	22	15	60	142	1.4×10^5	.171	6.6
117.	Am. Eagle Ealet	12.2	36'0"	2'0"	72	160	360	115	38	450	200	4.8×10^6	.150	1.2
118.	Ryson ST-100 Cloudster	100	57'8"	6'4"	213	1212	1650	150	42.5	895	438	3.7×10^7	.085	4.3
119.	Gossamer Condor	*.48	96'	12'6"	720	70	209	11.6	8.5	*17.4	139	2.8×10^4	.041	23*
120.	Gossamer Albatross	*.48	96'	12'6"	544	55	209	15.8	9.6	*30.3	154	5.2×10^4	.116	11*

* — Estimated

	HOME BUILT AIRPLANES	BHP (hp)	b (ft)	D_p (in)	S (ft^2)	W_e (lb)	W (lb)	V_{max} (mph)	$V_{s,0}$ (mph)	$R_{C,max}$ (fpm)	W_u (lb)	WV_{max}^2 (lb mph^2)	F_p (—)	$\frac{\sigma A_D}{\eta \phi}$ (ft^2)
121.	Christen Eagle II	200	19'11"	6'4"	125	1025	1578	184*	58	2100	553	5.3 x 10^7	.120	4.7
122.	Fisher — Van Norman P-51D	200				1450	1850	200		1500	400	7.4 x 10^7		3.7
123.	K&S Aircraft SA102.5 Cavalier	125	27'4"	6'	118	900	1500	185	40	1700	600	5.1 x 10^7	.194	2.9
124.	K&S Aircraft SA105 Cavalier	200	27'3"	6'	117	1000	1500	220	55	2500	500	7.3 x 10^7	.142	2.8
125.	Miller TM-5	65	25'	6'2"	100	805	1250	130	52	1030	445	2.1 x 10^7	.128	4.3
126.	Polliwagon	81	26'0"	4'8"	90	560	1100	168†	38	700	540	3.1 x 10^7	.101	2.5
127.	Quickie	18	16'8"		27.57	240	480	126	49	360	240	7.6 x 10^6	.089	1.3
128.	Quickie Aircraft Corporation Q2	64	16'8"		67	475	1000	180	64	800	525	3.2 x 10^7	.128	1.6
129.	Rutan Defiant	320	29'2"	5'9"	127.3	1610	2900	214#	75	1600	1290	1.3 x 10^8	.127	4.8
130.	Sequoia Falco F8.L	160	26'3"		107.5	1212	1808	212	62	1140	596	8.1 x 10^7	.091	2.5
131.	Thompson Boxmouth	55	24'	6'	400	350	550	45	30	500	200	1.1 x 10^6	.018	88.
132.	Thompson Metal Hurtant	65	23'		75	688	1200	120	70	600	512	1.7 x 10^7	.060	5.5
133.	Van's RV-4	150	23'	5'8"	110	890	1500	202	48	1650	610	6.1 x 10^7	.155	2.7
134.	Van's RV-6	150	24'		116	1000	1500	175	50		500	4.6 x 10^7		4.1
135.	Viking Dragonfly	45	22'	4'4"	97	590	1075	150		800	485	2.4 x 10^7		2.0
136.	Watson GW-1 Windwagon	30	18'	4'2"	54	273	485	135	45	450	212	8.8 x 10^6	.064	1.8
137.	Weedhopper JC-2-1	18.5	28'	3'9"	168	160	380	50	18	600	220	9.5 x 10^5	.138	21.7

* — 6,000 ft † — 75% Power, 8500 ft # — 70% Power

(Added in Revised Printing, 1981)

REFERENCES AND FURTHER READING

Abbott, I. A. and Von Doenhoff, A. E. (1949), *Theory of Wing Sections,* McGraw-Hill, New York. (Also Dover, New York, 1959).

Anon. (1962), *Basic Glider Criterion Handbook,* Federal Aviation Agency, Flight Standards Service, available from Superintendent of Documents, U.S. Government Printing Office, Washington, D.C. 20402.

Anon. (1974), *Federal Air Regulations, F.A.R. Part 23,* available from Superintendent of Documents, U.S. Government Printing Office, Washington, D.C. 20402.

Anderson, J.D., Jr. (1978), *Introduction to Flight,* McGraw-Hill, New York.

Bertin, J. J. and Smith, M. L. (1979), *Aerodynamics for Engineers,* Prentice-Hall, Inc., Englewood Cliffs, N. J. 07632.

Bruhn, E. F. (1945), *Analysis and Design of Airplane Structures,* Tri-State Offset Co., Cincinnati, Ohio.

Douglass, R. D. and Adams, D. P. (1947), *Elements of Nomography,* McGraw-Hill Book Co., Inc., New York.

Durand, W. F. (Ed.) (1934-6), *Aerodynamic Theory,* 6 Volume Set, Julius Springer, (Published by Dover, 1963)

Etkin, B. (1959), *Dynamics of Flight,* Wiley, New York.

Henderson, H. and Roemer, P. (1977), "A Study of Cruise Performance of the T-18," *Sport Aviation,* March, pp. 17-25.

Hoffman, R. (1935-1943), *Engineering for the Amateur Aircraft Builder,* Articles reprinted from *Popular Aviation and Flying.* Available from the Experimental Aircraft Association, P.O. Box 229, Hales Corners, Wisconsin 53130.

Hoerner, S. F. (1965), *Fluid Dynamic Drag,* Published by the author, P.O. Box 342, Bricktown, New Jersey 08723.

Kuethe, A. M. and Chow, C-Y (1976), *Foundations of Aerodynamics, Bases of Aerodynamic Design,* 3rd edition, Wiley, New York.

Liebeck, R. H. (1973), "A Class of Airfoils Designed for High Lift in Incompressible Flow," *Journal of Aircraft,* Vol. 10, No. 10. October, pp. 610-617.

Liebeck, R. H. and Ormsbee, A. I. (1970), "Optimization of Airfoils for Maximum Lift," *Journal of Aircraft,* Vol. 7, No. 5, Sept-Oct, pp. 408-415.

MacCready, P. B. (1978), "Flight on 0.33 Horsepower: The Gossamer Condor," AIAA Paper No. 78-308, Presented at AIAA 14th Annual Meeting and Technical Display, Washington, D.C., February 7-9, 1978.

Martin, J. F. (1977), "The Standard Sailplane — Parametric Studies," *Canadian Aeronautics and Space Journal,* Vol. 23, No. 1, Jan/Feb, pp. 1-18.

McMasters, J. H. (1974), "Low Speed Airfoil Bibliography," *Technical Soaring,* Vol. III, No. 4, pp 40-42.

Von Mises, R. (1945), *Theory of Flight,* McGraw-Hill, New York. (Also, Dover, New York, 1959).

Nicolai, L. M. (1975), *Fundamentals of Aircraft Design,* METS, Inc., Xenia, Ohio 45385.

Pazmany, L. (1963), *Light Airplane Design,* Published by the author, P. O. Box 10051, San Diego, California.

Perkins, C. D. and Hage, R. E. (1949), *Airplane Performance, Stability, and Control,* Wiley, New York.

Riegels, F. W. (1961), *Airfoil Sections,* Butterworth, London.

Roskam, J. (1971), *Methods for Estimating Drag Polars of Subsonic Airplanes,* published by the author, 519 Boulder, Lawrence, Kansas 66044.

Ross, F. W. (1948), "Unified Design for Performance and Safety Characteristics," *Aeronautical Engineering Review,* pp. 22-29, 84, May.

Schlichting, H. (1960), *Boundary Layer Theory,* translated by J. Kestin, McGraw-Hill, New York.

Sherwin, K. (1971), *Man Powered Flight,* Model and Applied Publications, Ltd., Argus Press Ltd., London.

Smith, A. M. O. (1975), "High-Lift Aerodynamics," *Journal of Aircraft,* Vol. 12, No. 6, June, pp. 501-530.

Sunderland, L. D. (1979), "Long Wings for Short Power," *Sport Aviation,* February, pp. 46-48.

Taylor, R. L. (1977), *Understanding Flying,* Delacorte Press/Eleanor Friede, New York.

Teichman, F. K. (1939), *Airplane Design Manual,* Pittman, New York.

Thurston, D. B. (1978), *Design for Flying,* McGraw-Hill, New York.

Weick, F. E. (1930), *Aircraft Propeller Design,* McGraw-Hill Book Company, New York.

Welch, W. A. (1972), *Custom Light Planes – A Design Guide,* Modern Aircraft Series, Sports Car Press, Ltd., New York.

Wendt, R. F. (1947), "A Method of Airplane Performance Calculation Applicable to Any Polar", *Journal of the Aeronautical Sciences,* Vol. 14, No. 4, pp. 243-250.

Wood, K. D. (1963), *Aerospace Vehicle Design.* Vol. I. Aircraft Design, Johnson Publishing Co., Boulder, Colorado.

Wortman, F. X. (1972), "Design of Airfoils with High Lift at Low and Medium Subsonic Mach Numbers," AGARD CP 102, Lisbon.

Wortman, F. X. (1973), "A Critical Review of the Physical Aspects of Airfoil Design at Low Mach Numbers," and "Airfoils with High Lift/Drag Ratio at a Reynolds Number of About One Million," in *Motorless Flight Research, 1972,* edited by James L. Nash-Webber, NASA CR-2315.

(Additional References on page 188)

INDEX

Abbreviations, 143
Absolute ceiling, 85-91
Advance ratio, 135
Airfoil, 16, 177-181
Airspeed, for best climb angle, 81
    cruise, 82-84
    for best rate-of-climb, 81
    maximum, 27-30, 80
    stall, 16
Altitude effects, 85-91
Angle of attack, 180-181
Aspect ratio, 21-24
Atmosphere, 85, 155-158
BASIC, 159
Bernoulli's equation, 132, 177
Boundary layer, 170-171
Brake horsepower, 17-28, 49-55
Ceiling, 85-91
Cessna 172, 105-108
Chord, 21-24
Circular cylinder, 172-173
Climb angle, 81, 130
Component drag, 169-175
Computer program, 159
Crawdad, 108-114
Critical Reynolds number, 170-173
Cruise speed, 82-84
Cubic equation, 134, 187
Delta wing, 22, 166-167
Density altitude, 85-91, 155-158
Density ratio, 185-186
Design procedure, 5-8
Dimensionless airspeed, 75-79, 127, 134
Drag analysis, 169-175

Drag area, 27-37, 169-175
Drag bucket, 167, 179
Drag coefficient, 121
Drag effects, 93-94
Dynamic pressure, 121
Effective aspect ratio, 21-24, 65-68, 124
Effective chord, 21-24, 45-47, 125
Effective span, 21-26, 43-44, 126
Effective span loading, 25-26, 33-37, 126-127
Efficiency factor, airplane (e), 165-168
    wing (e_w), 166
Elliptical wing, 21-24, 166
Equation of state, 185-186
Flaps, 93-94
Flow chart, 4
FORTRAN computer program, 159-163
Frontal area, 169
Fuselage, 165-168, 174
Gas constant, 185
Glide angle, 69, 120-122, 128
Gossamer Albatross, 99-102
Gossamer Condor, 99-102
Ground effect, 21, 168
Hang glider pilot, 173
Hang glider, powered, 102-105
Brake horsepower, 49-55, 133-137
Ideal rate-of-climb, 49-50, 75-81
Induced drag, 123-124, 165-168
Kinetic energy parameter, WV^2_{max}, 115-117
Laminar boundary layer, 170-171
Lift coefficient, 120
Lift coefficient for minimum sink, 45-47
    65-68, 125
Lift slope, 166, 181

Logarithmic graph paper, 75
Mach number, 57-59, 138
Man-powered aircraft, 99-102
Mass flow rate, 132
Maximum level flight speed, 27-30, 80
Maximum lift coefficient 15-18, 180
Maximum lift-to-drag (L/D) ratio, 43-44, 61-70, 128
Minimum drag, 33-37, 69-70, 128
Minimum power required for level flight, 33-41, 129
Minimum sink rate, 39-41, 61-63, 122, 125-127
NACA, 177-181
Nomogram 13, 149-151
Normally aspirated engine, 85-91
Performance data, 115-117
Performance rating parameter, F_p, 115-117
Performance relations, 138-139
Power-altitude factor, 85-86
Power coefficient, 135
Power effects, 96-98
Power setting, 82-83
Pressure, 185-186
Propeller, 51-59, 131-138
Propeller diameter, 51-59, 133, 135
Propeller disk area, 132-133
Propeller efficiency, 75-79, 132-134, 187
Propeller rotational speed, 57-59, 135, 138
Quicksilver, 102-105
Range, 82-84
Rate-of-climb, 75-81, 97-98
Rectangular wing, 21-24, 166
Relation ①, 15-18, 121
Relation ②, 19-20, 121
Relation ③, 21-24
Relation ④, 25-26, 126
Relation ⑤, 27-30
Relation ⑥, 31-32, 125, 169
Relation ⑦, 33-37, 126, 128-129
Relation ⑧, 39-41, 129
Relation ⑨, 43-44, 128
Relation ⑩, 45-47, 125
Relation ⑪, 49-50
Relation ⑫, 51-55, 134, 136-137
Relation ⑬, 57-59, 138
Relation ⑭, 61-63
Relation ⑮, 65-68, 125, 128
Relation ⑯, 69-70
Reynolds number, 178-179, 183-184
Section moment coefficient, 179
Service ceiling, 90-91
Sink rate, 39-41, 61-63, 75-80
Span, 21-26
Specific fuel consumption, 84
Speed of sound, 57, 138
Stall analysis, 16
Static thrust, 51-55, 136-137
Streamlining, 93-94
Struts, 172-173
Template 75-79
Thickness ratio, 172
Thorp T-18 Tiger, 9-11
Thrust horsepower, 27-30, 33-41, 129-130
Transitional boundary layer, 170-171
Transitional Reynolds number, 170-171, 184
Turbocharged engine, 85-91
Turbulent boundary layer, 170-171
Twin-engined aircraft, 96-97
Units 153-154
Useful load, 115
Viscosity, 183
Weight, 19-20, 25-26, 39-41, 49-50, 120-122
Weight effects, 92-93
Wetted area, 169, 174
Wing area, 19-24
Wing efficiency factor, e_w, 166
Wing loading, 15-20, 121
Wires, 172-173
Zero-lift drag coefficient, 31-32, 65-68, 123-124